CONFRONTING CLIMATE CHANGE

What are the manifest and likely future consequences of climate change? How will the world respond to the challenges of climate change in the twenty-first century? How should people think about confronting the politics of climate change?

In this highly accessible introduction to the predicted global impacts of climate change, Constance Lever-Tracy provides an authoritative guide to one of the most controversial issues facing the future of our planet. Discussing how the social and natural sciences must work together more effectively in confronting climate change, Lever-Tracy provides a sober, critical assessment of the politics of global warming and climate change.

By combining sociology, environmental studies and politics, *Confronting Climate Change* will serve as an introduction that will appeal to students and general readers alike.

Constance Lever-Tracy was Senior Lecturer in Sociology at Flinders University of South Australia, now retired. Her recent work includes editing the *Routledge Handbook of Climate Change and Society* (2010), and the entry for 'Global Warming' in the *International Encyclopaedia of the Social Sciences* (2008).

SHORTCUTS – *"Little Books on Big Issues"*

Shortcuts is a major new series of concise, accessible introductions to some of the major issues of our times. The series is developed as an A-to-Z coverage of emergent or new social, cultural and political phenomenon. Issues and topics covered range from Google to global finance, from climate change to the new capitalism, from Blogs to the future of books. Whilst the principal focus of **Shortcuts** is the relevance of current issues, topics, debates and thinkers to the social sciences and humanities, the books should also appeal to a wider audience seeking guidance on how to engage with today's leading social, political and philosophical debates.

Series Editor: Anthony Elliott is a social theorist, writer and Chair in the Department of Sociology at Flinders University, Australia. He is also Visiting Research Professor in the Department of Sociology at the Open University, UK, and Visiting Professor in the Department of Sociology at University College Dublin, Ireland. His writings have been published in sixteen languages, and he has written widely on, amongst other topics, identity, globalization, society, celebrity and mobilities.

Titles in the series:

Confronting Climate Change
Constance Lever-Tracy

Feelings
Stephen Frosh

CONFRONTING CLIMATE CHANGE

Constance Lever-Tracy

Routledge
Taylor & Francis Group

LONDON AND NEW YORK

First published 2011
by Routledge
2 Park Square, Milton Park, Abingdon, Oxon, OX14 4RN

Simultaneously published in the USA and Canada
by Routledge
270 Madison Avenue, New York, NY 10016

Routledge is an imprint of the Taylor & Francis Group, an informa business

Typeset in Bembo by Taylor & Francis Books
Printed and bound in Great Britain by
TJ International Ltd, Padstow, Cornwall

British Library Cataloguing in Publication Data
A catalogue record for this book is available from the British Library

Library of Congress Cataloging in Publication Data
Lever-Tracy, Constance.Confronting climate change / by
Constance Lever-Tracy.
    p. cm.1. Climatic changes–Social aspects. 2. Global
warming–Political aspects. 3. Climatic changes–Forecasting.
I. Title.
QC903.L48 2011
363.738'74–dc22   2010036130

ISBN: 978-0-415-57622-2 (hbk)
ISBN: 978-0-415-57623-9 (pbk
ISBN: 978-0-203-83008-6 (ebk)

CONTENTS

SERIES EDITOR'S PREFACE

In our own time, few matters are arguably as globally consequential as climate change. Climate change is riven by a conflict between scientific knowledge and everyday social practice. On the one hand, and whilst the science of climate change remains controversial, there is considerable agreement that as a result of higher levels of greenhouse gasses global temperatures are rising and will continue to rise, by perhaps as much as 5°C by 2100. On the other hand, people around the globe remain 'locked in' to the routines and rhythms of their oil-based mobile lives. In a striking treatment of the topic, Constance Lever-Tracey's lucid, sociological informed prose introduces the reader to a remarkably wide spectrum of issues arising from the governance of climate change. From scientific debates on global temperatures and changing climates to the vast literature on climate change technologies (such as innovations in wind, wave, tidal and geothermal energy), Lever-Tracy's *Confronting Climate Change* is a magnificent introduction and guide. A perfect shortcut indeed.

Anthony Elliott

PREFACE

This book discusses the issue of global warming and climate change, and its already manifest and likely future impacts on human life and society. It looks at ways to pre-empt the threats, by reducing our emissions through changed lifestyles and technological change. In the event of at least partial failure, it asks how and how far we can be prepared for the impacts and how we can try to survive them. Finally, it explores the obstacles and potential for effective initiatives, whether starting from the bottom up, at individual and local levels, or from the top down, through global targets and agreements. It assesses the chances of success, given the pressures to both co-operation and to conflict produced by shared dangers on the one hand, and by the rivalries of winners and losers, with vested and potential interests, on the other.

The known dangers and possible risks of climate change are unprecedented in human history in their scale, scope and complexity, perhaps even exceeding those of nuclear war. In the end, the outcomes of climate change, like its causes, will to a considerable extent depend on human actions. How, how fast and how deeply will we be able to reduce our impact on nature so that at least we can mitigate the worst outcomes? How effectively can we adapt to the changes now manifest, and prepare for the much more dangerous developments already in the pipeline?

The issue of how to confront climate change, and the degree of success or failure, will produce novel opportunities and pitfalls and quite new antagonisms and alliances that do not reproduce prior social, political or ideological fault lines. John Urry has suggested two likely scenarios for society, in a future of climate change. The first is a Hobbesian world of 'wild zones' and 'regional war-lordism', as inadequate pre-emptive action fails to curb the breakdown of 'mobility, energy and communication connections … a plummeting standard of living, a re-localisation of mobility patterns … and relatively weak imperial or national forms of governance'. The second is an Orwellian society, a 'digital panopticon' where reduced consumption is enforced by total control and monitoring of the strict carbon and mobility rations allowed to each individual.

This book tries to be more optimistic and devotes considerable space to the potential of various technological changes that might decarbonise our production and save us from the worst aspects of Urry's options. It concludes that the obstacles are economic and political rather than technical, and that success and failure would both involve major transformations of global society.

ABBREVIATIONS AND GLOSSARY

BP	British Petroleum
C	Centigrade
CCS	Carbon capture and storage
CO_2	Carbon dioxide
CSIRO	Commonwealth Scientific and Industrial Research Organisation
CST	Concentrated solar thermal
EU	European Union
Gen-1	First generation
Gen-2	Second generation
Gen-3	Third generation
Gen-4	Fourth generation
GM	General Motors
IFPRI	International Food Policy Research Institute
IPCC	Intergovernmental Panel on Climate Change
KW	Kilowatts
MW	Megawatts
NASA	National Aeronautics and Space Administration
NGO	Non-government organisation
OECD	Organisation for Economic Cooperation and Development
Ppm	Parts per million
REDD	Reducing emissions from deforestation and degradation
SNCF	French Railways
UK	United Kingdom
UN	United Nations
US	United States
Anthropogenic	Caused by humans
Mitigation	Reducing anthropogenic impacts

PART I
What do we know?

1

INTRODUCTION TO PART I

Social scientists have often assumed that nature can be taken as a given, stable background to the rapid changes in human history. Natural scientists have often been naive about, or paid little attention to, politics and society. Nature is, however, more complex and fragile and its seemingly predictable regularities are evolving more rapidly than we had thought. Society is more powerful, but also more ignorant and incompetent than we had often supposed. For centuries, social and natural sciences have separately studied their own domains. Their systems do indeed operate and change through discrete processes and laws. However, they also increasingly interact and mutually condition each other in real time, so that even a complete understanding of just the natural or just the social world would be inadequate.

Discovering global warming

Greenhouse gases act like a blanket, preventing the sun's heat dissipating and the earth freezing. Most natural sources, including water vapour, have a balancing cycle of emission and absorption that does not change much over time. Deforestation interferes with this cycle by reducing nature's capacity to reabsorb carbon dioxide, and human (anthropogenic) emissions have tended to accumulate at an accelerating rate since the Industrial Revolution. These include mainly carbon dioxide (CO_2) emitted by burning fossil fuels but also methane emitted by cattle belching and rotting vegetation (for example in rice cultivation), and by nitrous oxide emitted by artificial fertilisers in modern agriculture. Another effect of rising CO_2 in the atmosphere is that, as more of it is absorbed into the oceans, they become increasingly acid and less able to support many kinds of marine life.

While originally a specialised interest, global warming began to be publicly canvassed by climatologists in the 1970s. In 1979 the first World Climate Conference was held in Geneva, sponsored by the World Meteorological Organisation (WMO). Its place on global scientific and political agendas was further reinforced in 1985, at a conference in Austria of 89 climate researchers from twenty-three countries, participating as individuals. They unanimously forecast substantial warming, unambiguously attributable to human activities.

The basic principles of global warming and the role of industrial emissions had long been established. In 1822 the physicist J. B. Fourier had postulated that the earth was kept warm because 'air traps heat, as if under a pane of glass' and, over the next sixty years, others showed the critical role of rising levels of CO_2 and the other 'greenhouse gases'. In 1896 Arrhenius had reported his first systematic inquiry into the impacts on climate of the burning of fossil fuels by modern industry. In 1958 a young American scientist, David Keeling, began taking annual measurements of the levels of CO_2 in the atmosphere, in places far from direct sources of industrial pollution, finding continually rising levels.

Climate scientists had been increasingly concerned since the 1970s, but the meeting in Austria generated a wider realisation, within and beyond the scientific community as a whole, that significant changes could come in our lifetimes. A ball had been set rolling that led to increasing public and political concern. Growing numbers of scientists, from a wider range of disciplines and countries, began to accumulate and connect the dots of focused research.

The International Panel on Climate Change (IPCC), consisting of several hundred experts from around the world, was established by the United Nations in 1988, to monitor these accumulating findings and to evaluate the risks. Since then the greenhouse paradigm has guided targeted empirical research, multiplied supporting evidence, resolved anomalies and won near-unanimous peer endorsement. It has issued seven comprehensive assessments over twenty years. Each demonstrated greater confidence in the trends and rising concern about their implications. The numbers of serious sceptics were falling while others were discredited by revelations of dubious funding from fossil fuel industries or by their increasing dependence on global conspiracy theories.

Evidence and doubts

In April 2006 a NASA Goddard Institute oceanic study reported that the earth was holding on to more solar energy than it was emitting back into space. The Institute's director said: 'This energy imbalance is the "smoking gun" that we have been looking for'.

According to the British Meteorological Office, in 1998 global warming combined with the peak of the cyclical El Niño effect to produce the hottest average global temperature ever recorded. The first decade of the new millennium saw global average temperatures continuing to rise, albeit more slowly. No single year has yet reached the level of 1998 but the decade as a whole also broke all records, although there were localised colder winters in Europe in 2008 and 2009, due to Arctic winds from Siberia, (the result of the low point of the sunspot cycle): 'The linear warming trend over the past 50 years [0.13°C per decade] is nearly twice that for the past 100 years'. James Hansen of NASA believes that if the polar regions had been included in the calculations, records of global temperatures in 2005 would have exceeded those in 1998.

A US government report in June 2009 combined the findings of years of scientific reports and new data and revealed that average temperature in the US had risen by two degrees Fahrenheit over the previous fifty years. In Australia, the Bureau of Meteorology found that each decade since the 1940s was warmer than the preceding one. June 2010 was the 304th consecutive month in which the average global temperature, for land and sea, was above the twentieth-century average. The last month with below average global temperature had been in 1985. The second warmest year ever recorded until then was in 2009, with annual mean temperature 0.90°C above average, and by mid-2010 the year was shaping up to be even hotter.

The most dramatic and direct effect of the warming, visible to sailors as well as scientists, has been the melting of ice in polar regions. Just before the end of the last century, American researchers released ice-thickness data, gathered by nuclear submarines, that the Pentagon had been sitting on for ten years, which showed that over the previous forty years the ice depth in all regions of the Arctic Ocean had declined by some 40 per cent. Since then more careful and frequent measurement has found the melting speeding up. During the summer of 2008, researchers at Laval University in Canada found that five ice shelves in Canada's north had shrunk by a further 23 per cent, and a large chunk of the Ward Hunt Arctic Ice Shelf had broken free of the northern Canadian coast. The shelves that once covered 5,000 square kilometres were now only 10 per cent of that.

The Arctic Climate Research Centre estimated that by February 2009 there were 1.43 million square kilometres less Arctic sea ice (roughly twice the size of Texas) than at the same date in 1979. After cooling for two millennia the region had begun warming abruptly in the last century and was now the warmest it had been in 2,000 years. A NASA climate scientist told a gathering of fellow climate experts, at the end of 2007; 'The Arctic is often cited as the canary in the coalmine for climate warming ... and now the canary has died'. Later in 2011, the European Union's Cryosat-2 is expected to report more accurate and extensive radar measurements of ice thickness taken from space.

It was long believed that the Antarctic was less liable to warming and its ice more stable, because much thicker than the Arctic. A new analysis shows, however, that Antarctica has warmed 0.6 degrees over the last fifty years and it has become clear that its ice shelves are more fragile than originally thought. In the spring of 2009 the Wilkins Ice Shelf was reported to be in danger of imminent collapse, after the snapping of the ice bridge strip that connected the giant shelf to the Antarctic continent. In 1950 the strip had been almost 100 kilometres wide. Nine other shelves have receded or collapsed around the Antarctic peninsula in the past fifty years, often abruptly like Larsen A in 1995 or Larsen B in 2002. In total, about 25,000 square kilometres of ice shelves have been lost, changing maps of Antarctica. Ocean sediments indicate that some shelves had been in place for at least 10,000 years.

Warming increases evaporation and humidity. Global increases in humidity have now been confirmed by researchers and reported in *Nature* in 2007. Because water vapour is a greenhouse gas, this would amplify global warming in a feedback effect and would accentuate extreme weather events such as tropical cyclones and rainstorms. The observed pattern of humidity increases, in various parts of the world, resembles that projected by computer models of man-made global warming. The IPCC had said that such amplification was the largest 'positive feedback' mechanism they had identified.

Evidence of rising levels of CO_2 in the atmosphere is unambiguous. Despite difficulty in retaining research funding, Keeling persisted in his annual measurements and his curve shows the levels rising from 315 parts per million in 1958, to 378 by 2005, and reaching 387 in 2009, at the extreme end of predictions made only two years earlier. An eight-year European study, drilling Antarctic ice cores to measure the composition of the air bubbles trapped within, reported in 2005 that CO_2 levels in the atmosphere are now at least 30 per cent higher than any time in the last 65,000 years. The speed of the rise is unprecedented, from 280ppm before the Industrial Revolution to over 380 today. Long-term natural cycles have involved much smaller changes.

Meanwhile human greenhouse emissions continue to rise, from 1 per cent a year twenty years ago to 3 per cent now. Energy use in wealthy countries is still increasing, and newcomers such as China and South Africa are seeking to emulate their automobile and electricity use and production, with many and massive new power stations in the pipeline. New coal and oil sources are being sought, from Australia to the tar sands of Canada and under the Greenland Sea. Even if all the pledges made at the Copenhagen summit in 2009 were kept, they would be insufficient to keep global warming below the supposedly 'safe' increase of 2°C. Rises of 3°C by 2100 were likely.

The growing volume of research findings has demonstrated a speeding up of the processes. The rate at which the world's glaciers are shrinking has increased from an average of 30cm per year between 1980 and 1999, to 1.5m in 2006. Sea level rise is almost 80 per cent higher than some predictions. Global sea levels, which are estimated to have risen by 170mm over the twentieth century, have now been rising by an average of 3mm a year since the early 1990s and are now rising at double the rate projected earlier by the IPCC. Most authoritative scientific bodies predict that on present trends a point of no return will come within ten years, and that the world needs to cut emissions by 50 per cent or more by mid-century. Carbon dioxide levels in the increasingly acidic and saturated oceans, as measured in 2009, were at levels at the extreme end of 2007 predictions.

Climate science has also progressed through clarifying ambiguities and resolving anomalies. Attempts had been made by critics to discredit the methodology of Michael Mann's famous 'Hockey stick' graph, which showed average global temperatures over the last 1,000 years, with little variation for the first 900 and a sharp rise in the last century. After upward of a dozen replication studies, some using different statistical techniques and different combinations of proxy records, his results have been vindicated. A report in 2006, by the US National Academies of Science research council, supported much of his image of global warming history. There was, they argued, sufficient evidence from the tree rings, boreholes, retreating glaciers and other 'proxies' of past surface temperatures to say with a high level of confidence that the last few decades of the twentieth century were warmer than any comparable period for the previous 400 years. For periods before 1600, they found, there was not enough reliable data to be sure, but the Committee found the Mann team's conclusion that warming in the last few decades of the twentieth century was unprecedented over the last 1,000 years to be 'plausible'.

For long, measurements from satellites and balloons in the lower troposphere indicated cooling, contradicting those from the surface and the upper troposphere. In August 2005 a publication in *Science* of the findings of three independent studies described them as 'nails in the coffin' of the sceptics' case. These showed that it was faulty data, which failed to allow for satellite drift that lay behind the apparent anomaly.

Other doubts were raised by historical evidence of an exceptionally warm period in Britain in the Middle Ages (although not globally), probably caused by cyclical sunspot appearances rather than by human activity. Sunspots in 2007 reached the lowest level of their cycle while global warming has persisted. An initial report that some glaciers were retreating, was countered by a sceptical claim that others were advancing. A complete audit found the large majority were in retreat. Newspaper reports that polar bears were unable to access food and were drowning in melt water were attacked as fabrications. However, a subsequent study reported that the proportions of dead bears found to be fasting had risen from 10 per cent in 1986 to 29 per cent in 2006.

There remain many unknowns, in particular the effects of cleaning up pollution aerosols and the impacts of clouds. These latter are themselves increased by greater humidity, but may have contradictory results, either reflecting heat away or holding it close to the earth, depending on their height and structure. The speed of ocean acidification has come as a surprise to researchers. Research has, however, been advancing.

In late 2009, a number of accusations against the scientists and science of climate change were made by deniers and sceptics, and made media headlines. It seems plausible that these were politically motivated to undermine the Copenhagen Conference. They had trawled thousands of pages of IPCC documents and discovered a few errors of fact and procedure. In March 2010 the secretary-general of the United Nations asked the InterAcademy Council of the world's leading science academies to review these IPCC errors to make sure they were not repeated. He asserted there was no evidence to call into question the main findings of the IPCC. A similar conclusion was reported to the Dutch parliament in July, by the Netherlands Environmental Assessment Agency.

Sceptics had also hacked into emails of the Climate Research Unit at the University of East Anglia, claiming to have found evidence of data manipulation and of a conspiracy to keep negative findings secret and to silence critics. In April 2010 an independent committee, set up by Britain's House of Commons Science and Technology Committee, reported on the East Anglia scandal. They noted some minor errors but found no fault with the unit's scientific practice and no evidence of any impropriety. Similar conclusions have been reached by several enquiries commissioned by the university itself. The most extensive reported in July 2010 and exonerated the unit's researchers, whose 'rigour and integrity as scientists are not in doubt', and found no evidence of any malpractice or distortion of data. In June 2010 a Pennsylvania State University panel cleared Michael Mann (of 'hockey stick' fame) of allegations of misconduct that had arisen from the hacking of his email correspondence with British researchers at the University of East Anglia.

Growing scientific confidence in the existence and projections of anthropogenic warming has been unprecedented. In May 2001, sixteen of the world's national academies of science issued a statement, confirming that the IPCC should be seen as the world's most reliable source of scientific information on climate change, endorsing its conclusions and stating that doubts about them were 'not justified'. In July 2005 the heads of eleven influential national science academies (of Brazil, Canada, China, France, Germany, India, Italy, Japan, Russia, the UK and the USA), wrote to the G8 leaders warning that global climate change was 'a clear and increasing threat' and that they must act immediately. They outlined strong and long-term evidence 'from direct measurements of rising surface air temperatures and sub-surface ocean temperatures and from phenomena such as increases in average global sea levels, retreating glaciers and changes to many physical and biological systems'.

Public opinion is fickle and changeable. The so-called media beat up of the 'climate-gate' scandals and the loss of momentum after the stalled Copenhagen Climate summit in December 2009, contributed to renewed public scepticism either about the reality or extent of the warming trend or about its human (anthropogenic) causation.

No such decline has occurred among scientists themselves. Delegates from sixty-two academies to the April 2010 meeting of the InterAcademy Panel – the global network of the world's science academies – were asked what global issues concerned them most, 'looking ahead to 2020'. By a substantial margin climate change ranked first among scientists from both rich and poor countries. A study by Stanford University, published in *Proceedings of the National Academy of Sciences* in June 2010, surveyed the top 100 climate researchers in the world, and found that 97 per cent of them agreed with the IPCC's 2007 assessment of the reality and human causation of climate change.

In May 2010 a letter signed by 284 members of the US National Academy of Sciences was published in *Science* (DOI: 10.1126/science.328.5979.689) claiming that:

> There is compelling, comprehensive and consistent objective evidence that humans are changing the climate in ways that threaten our societies and the ecosystems on which we depend. Many recent assaults on climate science and, more disturbingly, on climate

scientists by climate change deniers are typically driven by special interests or dogma, not by an honest effort to provide an alternative theory that credibly satisfies the evidence. ...

(pp. 689–90)

It called for an end to 'the harassment of scientists by politicians seeking distractions to avoid taking action, and the outright lies being spread about them'.

2

KNOWNS AND UNKNOWNS

In the first years of the new century escalating, at least partly human-induced, climate change came dramatically closer to many. Evidence and warnings from respected natural scientific bodies featured almost daily in the media, as did accounts of record breaking heat, drought and wild fires, of exceptional hurricanes, melting ice, floods and rising sea levels that destroyed lives and livelihoods, spread tropical diseases beyond their prior range, threatened the survival of many species, and demonstrated the unreadiness of even the most powerful states.

While heatwaves, wild fires and drought fit most easily into public understanding of global warming, its connection with heavier rainfall or more violent storms or even exceptional cold spells requires explaining. Melting ice and rising or acidifying seas are often too remote or gradual to impress immediately.

Models used to predict the future incorporate all we know now, and are constantly corrected, revised and updated as our knowledge and the capacity of computers grows. Stephen Schneider writes:

> You draw conclusions based on what you think at the time, making your assumptions explicit; then you re-examine the assumptions in the light of new evidence; you recalculate ... That's

how science proceeds. A model provides the logical consequences of explicit assumptions. The real science is in how good the assumptions are – and that is where empirical testing and peer debate come in.

(p. 43)

Uncertainty about future trends derives partly from inadequate data and incomplete understanding. Much of it, however, is also a result of the large number of variables, whose extremely complex interaction may never be fully predictable. For these reasons a range of likely outcomes is usually given in scientific reports, with various probabilities attached.

Although the media like to report the most sensational predictions, scientists often prefer to avoid them, unless they are very well founded, although there are some exceptions. The egregious inclusion, in an IPCC report, of reference to a possible complete disappearance of all the Himalayan glaciers by as early as 2035, was experienced as extremely embarrassing and soon corrected. There was, however, no doubt about the fact of melting from some of these glaciers, confirmed by Chinese research and by Sherpa reports of the new dangers to Everest climbers posed by the melting ice. Partial evidence is sometimes acknowledged to be ambiguous. Earlier reports that the sea had washed over parts of low lying islands were later qualified by further study showing that the debris has sometimes extended beaches on the other side. Threats to the corals of the Barrier Reef result from an interacting combination of warming and weed growth, the latter caused by nutrient-rich water runoff from mainland agriculture.

Nonetheless it is clear that scientists have generally erred on the side of caution and conservatism in their future projections. Again and again recently, estimates have been revised upwards in the light of an observed speeding up of expected rates of change. What once seemed maverick predictions have often become mainstream. We have increasingly observed and understood the working of positive feedbacks and mutually reinforcing tendencies, but without precise measurements (often unavailable or impossible) their impact has often been excluded from models:

- The albedo effect is the increased reflection back of the sun's rays from white ice and snow. As these melt, and are replaced by darker water, the heat is absorbed instead, raising the temperatures locally on land and water. To this is attributed a part of the observation that polar regions and ice-capped mountain ranges have been warming faster than the rest of the globe.

- The oceans had been absorbing about one quarter of all carbon dioxide emissions. A recent study of all oceans, by Samar Khatiwala of Columbia University, found that they had accumulated one third more CO_2 than in the mid-1990s, and as a consequence were becoming more acidic, and thus less able to absorb any more. Scientists in Britain, using more than 90,000 measurements from merchant ships, found that the Atlantic Ocean was by 2005 soaking up only half the amount of carbon dioxide each year as in the mid-1990s. A variety of causes, including acidification, have led already to the collapse of some 30 per cent of ocean fisheries, which are producing less than 10 per cent of their original ability.

- In a hotter world there will be more evaporation and cloud cover. However, while some clouds at certain altitudes are known to be a greenhouse gas, holding heat in, other clouds may have the opposite effect, reflecting heat back away from the earth. Too little is known of cloud behaviour to calculate yet the net effect.

- Heat produces wild fires, which release the carbon stored in forests, leading to further warming. One estimate is that this already accounts for 20 per cent of all CO_2 emissions.

- Warming can also release other greenhouse gases. The thawing of the Siberian tundra, as the permafrost melts for the first time since the Ice Age, could dramatically increase the rate of global warming by releasing billions of tons of previously trapped methane (a very potent greenhouse gas). Methane hydrates are also stored on the sea bed not too far from the surface of continental shelves. After an eight-year halt, the amount of methane in the atmosphere started rising again in 2008. In March 2010 the journal *Science* reported huge leaks of methane from under the East Siberian Arctic shelf. Current methane concentrations in the Arctic were the highest for 400,000 years.

- Human responses to a warmer climate may reduce emissions when heating is turned down, but may increase it when air conditioning and the building of energy-intensive desalination plants increases.

It was because of such evidence, of rising emission levels and declining absorptive capacity, that the British economist Sir Nicholas Stern in 2009 doubled his estimate of the global cost of mitigating climate change, from 1 per cent to 2 per cent of global GDP per annum.

The precautionary principle states that where great or irreversible harm threatens, even a low probability must be taken very seriously. A study by Silva and Jenkins Smith, of how scientists in the USA and the EU interpret less-than-certain scientific findings, found contrasting responses to the dangers of low level radiation and of climate change. For the first danger, those who expressed greatest certainty about their own knowledge of the issue were least likely to fear major unexpected dangers or to advocate precautionary policies. For climate change the reverse was the case. Those who claimed the most knowledge were more conscious of the dangerous unknowns, and less willing to take even small risks that just might lead to irreversible major disasters.

3

MANIFEST VULNERABILITIES

Separating the impacts of long-term, global climate change from the noise of spatial and temporal weather variability, is not easy. As the twentieth century neared its close, most scientists had come to accept the reality and dangers of global warming, but a 1995 survey showed most still thought it a problem for a probably distant future. A decade later, few doubted its effects were already manifest, although the extent is still debated. For many people around the world their experience of a changing climate began to overtake the expected variations to which they were accustomed. Although many factors are likely to be involved in changing weather and some records are broken somewhere every year, the accumulating impacts became convincing for non-scientists as well.

A changing climate

Heat fire and drought

In Australia, the exceptional temperature of 2009 manifested itself in three unprecedented heatwaves: in southern Australia in the summer months of January and February; over most of the inland in the winter months; and again, in the spring, in the southeast. Over the last fifty years

Australia had experienced more record-breaking heat events than cold ones and this trend has been increasing over the last decade.

Water flows into the lower reaches of the major Murray-Darling river system fell drastically over several years, drying out and acidifying the stored groundwater and the lower lakes. Record-breaking rains in 2010 and 2011 have restored life to the system, but recurrence can only be prevented by cutting irrigation.

The 2003 European heatwave was, at that time, one of the hottest summers on record in Europe, especially in France. The heatwave led to health crises in several countries and combined with drought to create an agricultural crop shortfall in Southern Europe. The Centre for Disaster Research estimates 72,000 Europeans died as a result of the heatwave. Although many factors affect the weather, computer models, developed by experts at the UK Meteorological Office and Oxford University, showed the probability of such an event was at least doubled by man-made climate change.

Forest and bushfires have become increasingly common and devastating in different parts of the world. Although here too many factors are involved, the heat, dryness and the fiercer winds of climate change load the dice in favour of them happening. In June 2008 drought was declared in California, and towards the end of the year there were major fires. Governor Arnold Schwarzenegger declared a state of emergency, as 30,000 people were evacuated and 800 homes destroyed. In July and August 2009, unseasonal major fires struck again, devastating 525 square miles, and yet again later in the year. In Colombia at the end of 2009 an exceptionally short wet season was followed by a crisis of drought and fire.

In 2007 Australia's Climate Institute produced a paper, 'Bushfire Weather in Southeast Australia', which projected more extreme and cat-astrophic fire weather risks for this region of Australia, through each increment in global atmospheric warming. The paper did not yet, however, declare Sydney's 'Black Christmas' bushfires of late 2001, the Canberra bushfires of January 2003 or the 2003 and 2007 eastern Victorian bushfires to be directly caused by climate change, suggesting rather an unquantified 'mix of both natural variability and human-induced climate change'. Others blamed an increase of arsonist crimes, population movement or the protection of native vegetation which had interfered with preventative clearance.

In 2009, the heatwave in Victoria was followed by 'Black Saturday' – the worst bushfires in Australia's recorded history, with 78 communities affected, 200 houses destroyed and 173 dead. The firefighters union and the Climate Institute were no longer equivocal: 'These are the fires of climate change … Climate change is not just about warmer weather. It's about wilder weather. Climate change costs … climate change kills'.

In Greece the deadliest wild fires in recent history killed sixty-five people in 2007. In August 2009, fires fanned by gusting winds moved across the northern suburbs of Athens forcing thousands to flee, with mass evacuations including two children's hospitals.

Melting ice and rising seas

Initially the melting Arctic may have seemed of little direct relevance to most people's lives. However, concern has grown as the melting increased each year and then spread also to the Antarctic (contrary to expectations) and to glaciers around the world. Pictures of ice collapsing into the sea, and the now iconic images of hungry and perhaps drowning polar bears have filled television screens amid reports some were attacking domestic animals or digging up graves. The impacts of rising sea levels, changing weather patterns and shifting trade routes, as the sea-ice melting accelerated, began to be manifest.

By the end of the first decade of the twenty-first century, a previously ice-bound northeast summer passage to Japan is opening up. Rival national claims are heating up for access to the new shipping route and to the oil resources uncovered by the melting ice. A US Navy report points to the potential for conflict and warns that Russia is planning a new armed force to guard its Arctic coast. Loss of Arctic ice volume has been greater than loss of coverage, shrinking from 21,000 cubic kilometers in 1979 to 8,000 in 2009.

In February 2009 environmental ministers from sixteen nations, including Russia, China, India, the UK and the USA, visited the Antarctic (where 90 per cent of the world's ice is locked up, enough to raise the oceans by 70m if it ever all melted) to see the impact of accelerating global warming in receding or collapsing ice shelves.

An increase in rainfall has also accelerated the rate of melting and fracturing of glaciers along the Antarctic Peninsula, which in turn is adding to rising sea levels worldwide. The Sheldon Glacier has retreated 1.2 miles in the last twenty years. A report in *Geophysical Research Letters* in August 2009 describes how one of the largest glaciers in Antarctica appears to be thinning four times faster than ten years ago and nearly 90 per cent of the 400 glaciers along the peninsula are now in retreat.

In April 2010 there were reports of the splitting apart of the ice cap at the peak of the third highest mountain in Africa, in the Ruwenzori range in Uganda. Large losses have been measured in the Alps and Pyrenees mountain ranges in Europe, in Alaska and in Washington State in the US. In the Himalayas, accelerated melting is already contributing to the growth of dangerously swelling glacial lakes in Nepal, although there are suggestions that an increase of black soot in the atmosphere over Asia may also contribute to the unexpectedly rapid melting. The impact of glacier melt may change rapidly. While initially increasing water flows can contribute to the danger of flooding rivers and rising seas, water shortages and droughts can follow as the source is exhausted. In Bolivia the 18,000-year-old Chacaltaya glacier, on which tribal peoples and two cities depended for water, has almost disappeared as the retreat of Andean glaciers accelerates throughout Bolivia, Colombia, Ecuador and Peru.

The losses of sea ice and of floating ice shelves do not directly impact sea levels but they do remove the protective barrier from land-based ice, leaving it more exposed. Recently discovered underground lakes, expanded by melt water, have also been lubricating the underside of land-based ice and glaciers, speeding their break up and flow into the sea. Warmer and thus expanding sea water is another cause of coastal flooding.

Rising seas are among the most dangerous future threats of climate change (see Chapter 4) but so far measurement of their impact has been complicated by seasonal and local variability. A dryer climate in many regions has diminished flows from rivers, such as the Murray in Australia, which may somewhat slow down the sea rise. However, in low-lying Pacific island states, such as the Maldives, Kiribati and Tuvalu, and in fields around the Bay of Bengal, rising seas have already been driving internal climate change refugees from some coastal villages, where salt water is ruining fields and wells. Data from Bangladesh monitoring stations show average sea rises of 5mm a year for the last thirty years. In July 2010 researchers from the University of Colorado reported new measurements showing average rises of 13mm a year in parts of the Indian Ocean including the coastlines of the Bay of Bengal, the Arabian Sea, Sri Lanka, the Seychelles and Java. They attributed the rise mainly to anthropogenic changes in atmospheric and ocean circulation patterns.

Changing weather patterns

In 2007 the Australian Commonwealth Scientific and Industrial Research Organisation (CSIRO) noted that climate change seemed to be altering the balance of El Niño/La Niña weather cycles, decreasing the frequency of the former and thus extending the duration and frequency of droughts in some places, and of flooding rain in others, or of alternating extreme events in the same place. In some parts of Australia in 2010 prolonged drought was followed by exceptional heavy rains. The combination has brought on the worst plague of locusts in at least forty years.

In Africa, Uganda has suffered at least eight serious droughts in the past forty years, compared with only three during the previous sixty years. Lake Chad has lost 70 per cent of its water in the last thirty years, causing much hardship to those living on its borders in Nigeria, Cameroon, Chad and Niger. In February 2010 Mexico experienced floods from extreme rainfall in what is normally the dry season. In April in Brazil, intense rain triggered the most lethal landslides in Rio's history, leaving at least 224 dead, mainly in the hillside favela slums. European floods were the worst in 800 years, causing Hungary's store of toxic aluminium sludge to overflow, contaminating neighbouring countries.

An analysis of data on 925 rivers from 1948 to 2004, found climate change, through disrupted rain patterns and evaporation, had had an impact on about a third of them, with more than twice as many experiencing diminished flows as those with rising water levels. These rivers provided nearly three quarters of the world's supply of running water.

There are indications that severe hurricanes and cyclones have become more frequent. Early speculation about global warming as the cause became more vocal in the aftermath of Hurricane Katrina in 2005, when it was noted that the temperature in the Gulf of Mexico had been much warmer then normal.

Sir John Holmes, the UN relief coordinator, warned that 12 of the 13 major relief operations in 2007 were climate-related, and that this amounted to a climate-change 'mega disaster'. The insurance company Munich Re reported that in 2008 Cyclone Nargis in Burma (which scientists have directly linked to climate change) killed an estimated 130,000 people and devastated much of the low-lying Irrawaddy Delta. The most expensive single event in 2008 was Hurricane Ike (striking wealthy and well insured areas) which brought $30bn in losses. It was one of five major hurricanes in the North Atlantic over the year, which saw a total of sixteen tropical storms. In addition, roughly 1,700 tornadoes across the US caused several billion dollars of damage, as did periods of low-pressure weather activity in Europe.

The same year, the worst floods in fifty years hit the Indian state of Bihar, displacing 2 million people. According to Munich Re:

> The logic is clear: when temperatures increase there is more evaporation and the atmosphere has a greater capacity to absorb water vapour, with the result that its energy content is higher. The weather machine runs into top gear, bringing more intense severe weather events with corresponding effects in terms of losses.

> *(BBC)*

Cooling

Climate sceptics have pointed to the cold winter in Northern Europe in 2010 as disproof of global warming, but so long as global averages are rising, it just demonstrates disruption of previous weather systems. Some climatologists have explained how as wind configurations changed, the vortex of air spinning around the North Pole weakened and the cold air it had trapped there was released to flow south. Others note that as melted ice becomes cold water it spreads, cooling a wider area, like ice cubes in a glass of whiskey. More complex is the effect on the Gulf Stream, whose circular flow, from the North Sea to the Gulf of Mexico and back, brings warmth to Britain and Northern Europe, which is markedly warmer than the weather patterns on the other side of the Atlantic at the same latitudes. Melted ice diminishes the salinity of the sea water, making it less heavy and thus slowing down its sinking, and thus weakening the flow of the Gulf Stream in some years by as much as a third. This has resulted in warmer waters in the Gulf of Mexico while lessening the warming in northern climes. While the strength of the Gulf Stream varies from year to year, there was as yet no overall slowing trend between 2002 and 2009.

Manifest impacts

Impacts on the biosphere

A study in *Nature Geoscience* in 2007 reported that 'the tropics are expanding towards the poles, exceeding the worst case scenario of earlier models'. Changes in the Earth's physical and biological systems since at least 1970 were seen in regions which were known to be warming, it concluded. The researchers assembled a database of more than 29,500 records. Changes noted here and elsewhere include:

- Declining krill stocks around Antarctica
- The earlier arrival of migratory birds in Australia
- Earlier break-up of river ice in Mongolia
- Genetic shift in the pitcher plant mosquito in North America
- Declining productivity of Lake Tanganyika
- Earlier death of old growth trees in the western US
- Earlier annual appearance of cherry blossom in Japan
- A failure to spawn by millions of Pacific salmon
- Higher temperatures proving too much for many lizard species with declining numbers in France, Mexico and Africa and other places.
- Polar bears invading human settlements and interbreeding with grizzlies as habitats converge
- Declining shellfish stocks in an acidifying Arctic ocean
- The movement south of Australian east-coast climate zones, by 200km in sixty years, with warmer waters contributing to coral bleaching and fish off Australian coasts moving south.
- Stronger hurricanes destroying turtle nests in Florida.
- Decline in phytoplankton, the basis of the marine food chain.

Parmesan and Yohe reported in *Nature* in 2003, that 1,700 plant, animal and insect species had moved nearer to the poles, since the mid-twentieth century, at an average rate of about 6km per decade – unfortunately an adaptation too slow to match the changes in climate.

Hunger and disease

An Oxfam report, in mid-2009, reports that farmers around the world are already seeing changes in weather patterns which are leading to increased ill-health, hunger and poverty. Oxfam staff in fifteen countries collected records from communities and observed that: 'Once-distinct seasons are shifting and the rains are disappearing. Poor farmers from Bangladesh to Uganda and Nicaragua, no longer able to rely on centuries of farming experience, are facing failed harvest after failed harvest.' The evidence of the impact of changing weather patterns is anecdotal but the results are striking because of the 'extraordinary consistency' they show across the world. Farmers were all saying very similar things: the seasons are changing. Moderate, temperate seasons are shrinking and vanishing. Seasons are becoming hotter and drier, rainy seasons shorter and more violent.

Rising temperatures have also extended the range of malarial mosquitoes in Africa, into more densely occupied highlands previously free of them. In East Africa flooding created a breeding ground for mosquitoes carrying Rift Valley Fever. Dengue Fever is spreading south from Northern parts of Australia and has recently re-invaded the US.

Conclusion

Natural variability and the contradictory impacts of interacting trends can make it hard to attribute causation. At any date there will be unusual or even record-breaking weather events somewhere. Simultaneous exceptional periods of cold and heat, flooding rains and droughts in different places, may seem to make no coherent sense. November 2009 saw both the hottest day ever recorded in Adelaide and the wettest in Britain. The start of 2010 saw the worst snow storm ever recorded in Maryland, at the same time as Vancouver experienced the hottest January, threatening the winter Olympics. What they have in common is their deviation from previous patterns and their tendency to extremes. To assess the reality and extent of climate change we must look to longer-term and to global trends which go beyond direct experience and require huge efforts of measurement and analysis.

Global assessments such as those described in the previous chapters, far from being alarmist, have often tended to underestimate, out of caution, and to lag behind unexpected and accelerating changes that can find us unprepared when they become manifest. For example David Karoly, an IPCC lead author, has explained that publishing deadlines meant that their 2007 full fourth assessment report could only take account of peer-reviewed research published by about mid-2006, with these articles in turn involving research done no more recently than 2005. The report did not include recent observed changes in a number of climate variables that had been at or above the upper range of IPCC assessments. These included findings that sea-level rise in recent decades was 'at the upper bound' of what had been anticipated by the IPCC, thinning of Arctic sea ice was already 'consistent with 2030 climate model simulations' and CO_2 emissions were growing faster than any of the IPCC emission scenarios.

4

FUTURE RISKS

Projecting future risks is complex and must allow for many uncertainties, but according to scientists the credible long-term risks from unmitigated climate change are unprecedented in human history. The latest IPCC report, in 2009, estimated that under a 'business as usual' scenario, global temperatures could be 7°C higher by the end of the century, with catastrophic consequences. Fear of such outcomes has spread beyond the ranks of scientists and environmentalists. Australian economist Ross Garnaut, for example, has argued that even a three-degree rise could destroy the Great Barrier Reef and the tropical rainforests, cause widespread desertification, a mass extinction, and ultimately spark a huge sea-level rise.

The Global Climate Risk Index, presented to the Copenhagen climate change conference in December 2009, ranked Bangladesh as the most vulnerable country, followed in the top ten by Myanmar, Honduras, Vietnam, Nicaragua, Haiti, India, the Dominican Republic, Philippines and China. Italy, Portugal, Spain and the US qualified for inclusion in the top twenty countries at most risk.

Future climate

Heat, fire and drought

A study published in 2007 by the United Nations University, with the co-operation of more than 200 experts from twenty-five countries, concluded that climate change was making desertification 'the greatest environmental challenge of our times'. Desertification could displace up to 50 million people over the next decade, driven from their homes by encroaching deserts, particularly in sub-Saharan Africa and Central Asia. A British study, cited by the IPCC, estimated that by the 2020s 'between 75 and 250 million people [in Africa] are projected to be exposed to an increase in water stress due to climate change', although other parts of the continent would experience higher rainfall and greater access to water.

In 2008 Adelaide experienced fifteen continuous days over 35 degrees, the longest ever recorded for an Australian capital city. Meteorologists consider that a greater frequency and intensity of such heatwaves is associated with changes to the La Niña weather system, brought on by global warming, and are likely to recur. Models suggest global warming could bring temperature rises as high as 6°C for Australia this century, if global emissions continue unabated. Major scientific bodies in Australia have predicted the likelihood of bushfire dangers rising threefold by 2050.

Scientists have warned droughts could lay waste to tracts of US and that the UK 'must plan' for a warmer future, with wetter winters, drier summers and warmer weather all year. Global water demand would outstrip supply by 40 per cent by 2030 unless action was taken to stop climate change or find new water sources for thirsty cities.

Model results indicate that in a greenhouse-warmed world the rainfall belts in both hemispheres will move pole-ward. Drying is thus predicted to occur in 'Mediterranean-type' climates (climate with winter rainfall maxima) in southern Europe, California, southern Africa and Australia, with floods in the northern parts.

Melting ice and rising seas

The Arctic could be ice-free in summer by 2013, claims a new US scientific study. Rising seas are directly related to increasing temperatures and the rate at which ice melts. Extrapolation based on IPCC scenarios suggests sea levels would rise between 0.5 and 1.4 metres by the end of the century. Even a rise of 1m would threaten 600 million people and many major cities with inundation, but many scientists consider this an underestimate. Pfeffer suggests a range of scenarios from 80cm to 2m, but even this depends on the rest of West Antarctica, behind the very thick and cold Ross and Ronne ice shelves, staying put. Their break up seems unlikely, but a change in wind patterns or warm currents could alter this expectation. A recent study by Oxford University researchers suggests the Pine Island Antarctic glacier has already passed a tipping point and is poised to lose enough ice to raise by itself global sea levels by 24cm within 100 years, and possibly double that amount.

Disrupted weather patterns

The Indian Ocean tsunami, the worst disaster of this century's first decade, was caused by an underwater earthquake most unlikely to be related to climate change. However, the OECD reports on comparable storm and sea surge threats caused by warming, humidifying and shifting air currents and changing patterns of rainfall. These are likely to cause more frequent or more intense storms and floods, often in places previously unaffected and unprepared. The Indian Centre for Science and Environment and the Australian CSIRO also warned in 2008 of stronger cyclones and rising sea surges that would result from global warming.

Bureau of Meteorology models have suggested that changed weather patterns in Australia were likely to continue and accentuate. Along with the higher temperatures, rainfall would decrease further in the southern states, but would increase in the north. Researchers document a decrease in growth rates of storm track modes and effectively a poleward movement of mid-latitude storm tracks in the Southern Hemisphere post-1975. A Canadian expert says the impact of melting ice shelves on cyclones and floods will be first felt in higher northern latitudes. The Irish Climate Research and Analysis unit predicts a wetter climate with more rain over the next fifty years. There are indications that severe thunderstorms in eastern and southern US could double by 2100.

Future impacts

Biosphere

The dates of the seasons, and the habitat boundaries of animals, plants and microbes will shift. A 2009 report shows one third of all species are now threatened with extinction by multiple dangers, with climate change high on the list. The already fragile biosphere will take time to adapt to major changes in climate. By then it could be too late for many species. Ecosystems are interdependent, and rapid changes are likely to be transmitted widely to plant, animal and microbial life, as well as to agriculture and to human life.

The indirect effects of such changes can upset natural systems in ways more far-reaching than the direct ones. Over the last few decades, for example, the arrival of spring came progressively earlier in the UK. A major study, reported in *Global Change Biology*, examined 25,000 records of 726 marine, terrestrial and fresh water species, and found that different levels of the food chain were being thrown out of synchronisation, as species adapted at different rates from each other. If this were to continue, they concluded, key ecosystems and the survival and breeding requirements of many interdependent species could be disrupted.

Even if achieved, the emissions targets being discussed in UN conferences such as Copenhagen, would probably be insufficient to save the world's coral reefs, upon which many smaller fish species depend, and Australia's Barrier reef could be gone within thirty years. The remaining Bengal tigers are in danger of extinction as seas rise in the wetlands of the bay. Disintegrating ice and warming seas will threaten polar bears and king penguins. Even a couple of degrees temperature rise could threaten the survival of birds in desert areas.

Impact on humans

In May 2009, Kofi Annan's UN think tank, the Global Humanitarian Forum, released the first comprehensive study of the impact on humans of global warming. It projected that heatwaves, floods, storms and forest fires would be responsible for as many as half a million deaths a year by 2030, making it the greatest humanitarian challenge the world faced.

If emissions were not brought under control within twenty-five years, the report stated:

- 310 million more people would suffer adverse health consequences related to temperature increases
- 20 million more would fall into poverty
- 75 million more would be displaced.

Rising seas threaten highly populated and intensively cultivated low-lying areas. The IPCC report, in 2007, named Bangladesh, Vietnam and Egypt, on the Ganges, Mekong and Nile deltas, as the three countries most vulnerable to rising waters. Over 20 million people in Bangladesh are considered likely to be at risk of flooding and salination in coming decades, with vast tracts likely to be inundated every monsoon season, destroying their staple rice cultivation. On the other side of the Bay of Bengal, India is already considering provision for evacuating 70,000 people from the Sunderbans over the next five years. Even just a 50cm sea rise could displace 10 per cent of the population of Egypt. China's Pearl River Delta will also risk more flooding and salination, as seas rise. Twenty-six per cent of the territory of the Netherlands is at early risk of flooding because it lies below sea level.

How bad could it get?

The material, in this chapter and the last, demonstrates how often the accelerating speed of climate change has been underestimated. As our knowledge and understanding of the processes have advanced, the incalculable dangers have come to be seen as more threatening and imminent. What seemed sensationalist and unlikely a decade ago, no longer seems impossible. The IPCC hardened its stance, during 2009, on the possibility of 'abrupt and irreversible' 'large scale singularities', including collapsing ice sheets (uncontrollable once started), Gulf Stream shutdown and runaway warming which might raise sea levels every year for thousands of years, unstoppably.

James Hansen questions where the notion of a 'safe' global' temperature increase of up to two degrees comes from. He points out that the last time the earth was two or three degrees warmer than today was about

3 million years ago, when sea levels were 25m higher than today (covering land where currently about a billion people live) and Florida was under water.

Dave Stainforth, of the Tyndall Centre, and others suggested in 2007 an extreme but not impossible long-term scenario in which a collapse of the global ocean thermohaline circulation system could shut down the Gulf Stream, which protects Britain and Northern Europe from freezing temperatures and a new ice age, 'switch off' the Asian monsoon and create a warmer Southern Ocean, destabilising the West Antarctic ice sheet. This could eventually cause a further 7m rise in sea levels and perhaps a near permanent El Niño in the Pacific, which would hasten the burning of the Amazon. If all the ice in the Antarctic melted it would raise the oceans by 70m. The hot, lifeless planet Venus once resembled the Earth, before a runaway greenhouse effect boiled its water to vapour and its carbon to carbon dioxide.

Some recent studies of the distant past even indicate that rapid catastrophic change, on a human time scale, has already occurred before. It is not impossible that mutually reinforcing feedbacks could lead rapidly to irreversible 'tipping points'. Researchers at the University of Mexico recently reported, in *Nature*, evidence of a 2–3m rise in sea levels, over just 50–100 years, seemingly caused by ice sheet instability during the last interglacial period. They warn that today's ice sheets could deteriorate, owing to anthropogenic global warming, and initiate a rise of similar magnitude by the end of this century. In November 2009, scientists reported evidence of a huge Canadian lake suddenly overflowing and spilling into the Atlantic, 12,800 years ago. The impact of the fresh water slowed the Gulf Stream and 'within weeks' an ice age that lasted around 1,300 years began to engulf Europe.

5

CONFRONTING THE RISKS

The risks of climate change, as described in previous chapters, combine varied localised causes and impacts with broad global trends, and they present us with sudden, unpredictable disruptions as well as with gradual future tendencies that may be irreversible for centuries or millennia. While scientists focus mainly on the broad parameters of climate change, and are increasingly convinced of their reality, the direct experience of most people is much more diverse and discontinuous and the public and media are often diverted and distracted by short-term and local weather events. If we are to recognise, adapt to and contain the threats, we must act with local knowledge and extreme urgency, but also in ways that lead to overall long-term transformations of our societies and economies.

Tackling these contradictory needs together is hard. It seems clear that only a co-operative worldwide effort has a chance of success, yet a past history of clashing interests, inequalities and conflicts makes this difficult. How fast we move, how far and in what direction will affect different groups in different ways, producing also new winners and losers, or at least different levels of loss. Nonetheless, we might take some hope from two elements. First, perhaps paradoxically, is the fundamental unpredictability of the threat, that makes us all vulnerable, and that can undermine self-seeking calculations of distinct interests. Second is the globally interconnected and networked nature of our world today, which extends the possibility of distant knowledge and empathy.

Obstacles and opportunities for united action

Local experience versus the big picture

Changes in the weather have some direct impact on every person every day, but temporary and localised variability can mask the overall long-term global climate trends and risks. As experienced so far, warming itself has been generally small, unevenly distributed and sometimes interrupted, with a disproportionate amount in the polar regions, far from most people's daily lives. In some places and times there has actually been cooling. For example the year 2008 was not as hot as the preceding ones; the annual progressive bleaching of the Great Barrier Reef was interrupted in 2009 by a deep monsoonal trough and local cyclones, which swept away the warming ambient water; droughts have in many places been interrupted by flooding rains.

There is thus a large difference between weather, of which there is personal and local experience, and climate, which can only be conceptualised by a process of abstraction, combining into averages thousands of partial and often contradictory pieces of the jigsaw, from all parts of the globe. Any long-term or large-scale pattern or trend only emerges from records kept over many years, and from older evidence in markers concealed beneath the ice or in stalactites or ancient tree rings or the fossil record, or in the depths of the sea or the upper atmosphere. Despite this complexity, scientists find it much easier to be sure about large-scale, long-term trends than about short-term, local weather forecasts. To get these latter reliably right, it has been suggested, we would need to multiply our computing power a thousandfold.

For most people, however, direct local experience trumps theories, models and broad averages. Everyone knows the short-term variability of the weather, and in some place or another, records are broken every season.

When the devastation by Hurricane Katrina of the iconic city of New Orleans was shown on television around the world, a possible connection with global warming was not at first mentioned by the victims or the reporters on the spot. Over the following months, however, the media increasingly gave attention to scientists' measurements of changes not visible to local people. These indicated a possible link to the hotter waters in the Gulf of Mexico, themselves a product of the slowing down of the Gulf Stream's cold water flow from the north.

Confidence in such distant and abstract measurements depends on trust in the competence and honesty of scientists, and can rapidly dissipate, the more so as they often fail to provide accurate short-term and local weather forecasts. Such loss of confidence was demonstrated in opinion polls in the US and in Britain, in the winter of 2010, as exceptional snowfall in the Americas and northern Europe coincided with some minor revelations about scientific irregularities, with doubts fanned by vocal sceptics. A survey in the US in 2007 found 84 per cent believed the planet was warming, but in 2010 only 74 per cent remained convinced of this. In the UK in 2005, 91 per cent had believed this but only 78 per cent in 2010. Another UK poll in February 2010 showing only 26 per cent believing human causation of climate change was 'an established scientific fact' compared with 41 per cent three months earlier. However, the BP oil spill in the Gulf of Mexico seems to be reawakening public concern about our insecurity in the face of nature and our capacity to affect it.

A difference between the time horizons of concern to scientists and to other people is perhaps even more important than the spatial ones. Initially scientists did not expect noticeable impacts for decades or even centuries. Even now, manifest effects have only had a major impact in a few places, while it is usually the more distant projections that range from the severe to the catastrophic. It is often claimed that people naturally have limited ability to focus on such a distant future. It is doubtful if such short-sightedness is embedded in our genes, but the last decades have produced a cultural change with a speeding up of time horizons for economic and political calculations and ever more rapid changes of life-style and technology.

Undoubtedly the current speeding up of the actual manifestations, as well as the foreshortened projections, of climate change impacts have pushed the issue up the public and political agenda and increased the sense of urgency to take action. It is also possible that a significant cultural transformation in attitudes to future dangers and responsibilities could develop, especially among a new generation of young people. It is, however, also possible that the gulf between science and experience, between climate and weather, could widen again, as happened in the winter of 2010.

Losers versus winners

The obvious and immediate losers from the direct impacts of climate change are clear. Some of those vulnerable to heat, rising seas and salinity, flooding rains and storms, or to drought and water shortages, have already experienced the early impacts. Low-lying Pacific islanders, African herders and Bangladeshi farmers may by now see the writing on the wall. Less immediately, catastrophe may be waiting just down the tracks for those in the Nile Delta or dependent on rivers fed by Andean glaciers.

Mass extinctions have been predicted for many species, under pressure from the expansion of human society, and the process will be accentuated and accelerated by climate change. Seas, acidifying as they absorb more carbon dioxide, are already endangering many kinds of ocean life and those who depend on them. Beyond these, the speed and novelty of apparently harmless changes will pose problems for societies and species that lack the resources to adapt fast enough.

Since all greenhouse gases combine to circulate the whole globe, there is no direct causal link or proportionality between the emissions made by one nation or group and their vulnerability to the consequences. The effect of emissions from factories, cars or decaying chopped up forests can strike distant places, often unpredictably. A United Nations Human Development Report in 2007 warned that it was the poorest nations, least responsible for past emissions, and with the weakest voices, who risked a downward spiral of malnutrition, water scarcity and ecological damage from climate change. The 2009 report of the UN's Global Humanitarian Forum estimated that, of the twenty-eight countries at extreme risk from climate change, twenty-two were in Africa.

The situation is, however, more complicated. Natural processes know neither justice nor malice. The Victorian bushfires took 173 Australian lives and, in 2003, the European heatwave killed 72,000. Coal and oil companies have continued to make high profits, but insurance companies have already faced massive claims, and the fashionable beachside resorts and gated communities of Florida or Queensland's Gold Coast, although in great danger, may now be uninsurable. If the seas rise beyond a certain height, major coastal cities such as London, New York and Sydney will be inundated. If the Gulf Stream fails, Britain will be plunged into a climate like that of northern Canada, more appropriate to its latitude. It is true that the wealthy have more resources with which to adapt and defend themselves, but their survival and coping skills may often be less than those of the poor.

There will of course also be winners, but surely far fewer than the losers, if only because unexpected knock-on effects will have wider impacts. As deltas and islands are flooded, refugees will hammer at the gates of those not directly affected. The global trade and financial systems will be undermined, the more so because of the unpredictable nature and targets of tipping points and catastrophes.

Still, some will surely gain. Within a decade, according to Norway's foreign minister, a new ice-free summer passage for ships will open, around Russia's Arctic coast and across the North Pole, cutting tanker time between the Netherlands and Japan by 40 per cent. Many in colder climes will welcome the warmth, even if interrupted by floods and storms. Some have estimated that here cold normally kills more than heat, and lives may on balance be saved. While much of the world's agriculture may be trashed, longer growing seasons may increase the now more valuable yields in other places.

Conflict or co-operation

Climate change threatens everyone, but in different and unpredictable ways. It is caused, to varying degrees, by all people on earth, by their industry, transport, agriculture and land and forest clearing practices. While few in poorer countries bear any responsibility for the past accumulation of emissions, very populous countries in Asia and South America are responsible now for the fastest increases in greenhouse gases in the atmosphere. China's emissions have overtaken those of the USA.

Long-established conflicts and sources of mistrust may, however, distract from the unprecedented nature of the crisis and from the need for a shared response. Racism caused the fleeing blacks of New Orleans to be either ignored or halted and even shot at road blocks, seen as dangerous looters and marauders, rather than as victims of the hurricane, who needed help to evacuate. Distant empathy for the city sometimes exceeded local solidarity, in a region still split by ancient race and class prejudices.

Poverty and conflict in Africa are in part products of previous causes and in part exacerbated by early manifestations of climate change. These may be seen as just an endless continuation of dysfunction, or alternatively as another manifestation of the legacy of colonialism or of continuing imperialist plots. Co-operation between the two largest emitters, the US and China, is hampered by the longstanding fear of job losses to China, and by the Chinese sense that their development and rise out of poverty are being blocked by America. Each suspects the motivations of the other in their rivalry for great power status.

Climate change is also, however, likely to throw into disarray prior hierarchies and certainties, promoting new interests and alliances but also new conflicts, including the possibility of water wars and violence against refugees. Rival national claims are heating up for access to new shipping routes and to the oil resources uncovered by the melting ice. A US Navy report warns that Russia is planning a new armed force to guard its Arctic coast.

While danger may foster xenophobia, the common and unpredictable nature of our present and future vulnerability may also promote empathy and solidarity, such as was seen after the Indian Ocean tsunami and the earthquake in Haiti. In both cases media publicity was important. In the case of Haiti the internet facilitated the spread of information and of multiple decentralised responses. In both cases a motivator of wider co-operation was the presence among the victims of people from many other parts of the world – holidaymakers, NGO and UN staff, and others.

At least as important was the presence, around the world, of people with close personal and familial links to those struck by the disaster, including expatriate workers, students and emigrants. The Haitian community in New York rallied and lobbied. Some of them were able to return bringing skills and resources. Many climate refugees of the future will find in such communities an established base, ready to listen and to receive them, and perhaps to hide and protect them if necessary.

In our globalised world, the ripples from a local impact can extend a long way. Both the US and China are among the twenty nations most vulnerable to future climate risks, and together they have the weight to slow or halt the warming. Perhaps the greatest force for co-operation is the globally networked community of scientists, now significantly radicalised by the issue and by the attacks on their professional credibility. They share a culture that understands and focuses on the long-term, systemic and abstract attributes of climate change rather than on the shifting vagaries of the weather. The issue has seen them becoming vocal and united to an unprecedented degree.

Conclusion

Our vulnerability to anthropogenic climate change derives from inter-connected and cascading systemic causes, ultimately derived from the totality of human actions across the globe, over the last 250 years or so. Scientists from around the world have produced a collective under-standing of causation and operation, which despite many gaps and ambiguities is becoming more demonstrable and coherent all the time, leading to an unprecedented degree of near unanimity by the experts.

However, effective public recognition and response to climate change, which for scientists is a global problem requiring worldwide action, may clash with the very local experiences of changing weather. While there may even be some winners, the vulnerability of the many losers will be manifest in different ways and degrees and to different timetables, and be just as likely to produce conflict as co-operation.

However, unpredictable and cascading local events have distant reper-cussions. Fires in Russia can spike world food prices. In today's inter-connected world we can observe distant experiences in real time in our own living rooms. So long as we do not categorise them with the fictions of advertising or disaster movies, these may motivate a more global con-sciousness than was available in earlier periods. It was the sight of crum-bling Arctic ice on the screen which first aroused the concern of the author of this book.

The global media beam pictures of climate disasters into every living room and give the question of climate change a higher profile than ever before. Global movements of people, for migration and for temporary work and holiday visits, have ensured that any localised event will evoke calls for solidarity and aid far beyond its borders.Even at the deadlocked conference in Copenhagen, the world's political leaders jointly agreed about the nature of the problem and a threshold of two degrees warming that must not be crossed.

PART II
What can we do?

6
INTRODUCTION TO PART II

Introduction

At the Copenhagen conference, in December 2009, there was agreement that the accumulation of CO_2 in the atmosphere had to be kept, at most, to a maximum of 450 parts per million (ppm) if there was to be even a 50/50 chance of holding global temperature increases to an 'acceptable' 2°C. While there was some claim that two degrees of warming could avoid the worst impacts, many scientists and protesters outside the venue, as well as representatives of small island nations, argued that a goal of one and a half degrees was necessary, requiring an actual reduction in CO_2 to 350ppm.

The Kyoto protocol's emission targets, agreed in 1997, had not been met. Emissions in developed countries were continuing to rise, prompted by advertising and ever more rapidly changing fashions, only briefly interrupted by the global financial crisis in 2009. This hiatus was balanced by continually rising emissions in the developing world, especially China and India, as their populations and their new capitalists and middle classes also laid claim to the same material goods which the developed had achieved. Energy consumption in Asia grew by 70 per cent in the ten years to 2008. China has been for some time the world's leader in total greenhouse emissions, although still well down per head of population, and recently became also the world's largest consumer of energy.

In the end there was deadlock in Copenhagen on targets and carbon reduction measures to achieve even the two degrees goal, and little progress was made in the nine months after the Copenhagen conference. Meanwhile the globe continues to get hotter and the brief window of opportunity for effective mitigation is closing. We must, however, persist in seeking such measures because any level of mitigation, however inadequate, is preferable to 'business as usual', which would lead to temperature rises of 7°C or more, by the end of the century, and perhaps yet more, unstoppably, thereafter.

There are two main routes proposed for the needed transformation. Proponents of each argue the impossibility of the other, yet clearly both are needed. The first involves the world moving to a sustainable, no-growth economy by doing things differently with less waste and much reduced consumption of unnecessary material goods. The second route is the rapid development and adoption of alternative energy technologies.

Insofar as much damage is already happening and more is now unavoidably in the pipeline, we must also focus on ways of preparing ourselves and adapting to the likely dangers that our delayed mitigation will produce. The following chapters will discuss various often-competing proposals for confronting climate change.

What can we do?

Short- and long-term responses

Some changes are unfolding even without global agreement. Efficiency may be ignored by the short sighted, but its cost saving benefits can operate through market forces, especially if the price of energy is rising as supplies dwindle. Some shift in consumer demand from material goods to services follows naturally from rising prosperity and also from an ageing population. Early hopes that these might suffice have, however, disappointed.

As soon as economic recovery began, the prior rate of global emissions resumed. The interruption, based on belt tightening not innovation, proved unsustainable. New sources of fossil fuels, in the tar sands of

Canada or under the oceans, are being explored and coal- and petrol-burning power stations, vehicles and aircraft are multiplying. By the end of the first decade of the century, CO_2 concentrations in the atmosphere had passed 383ppm, and were rising, compared with a level in pre-industrial times of 265ppm.

Short- and long-term mitigation can work together, as the second builds on the earlier experience, knowledge and infrastructure. They may also, however, pull in opposite directions. A partial remedy, with one-off or diminishing returns, which merely slows down the growth of our dangerous emissions for a time, without the capacity eventually to replace them, will by itself do no more than postpone the inevitable and may promote complacency. On the other hand many long-term remedies will involve short-term emissions spikes, as cement for hydro dams, glass for insulation or solar panels and iron for rail tracks or windmills is manufactured in greater quantities. In a world of scarce resources and rival interests, hard choices must be carefully made and then followed through consistently.

The impacts of climate change will strike in different ways and at different times. The immediate temptation is to develop a bunker mentality, exclude outsiders and refugees and fight for scarce resources. Yet just as long-term global solutions are needed to mitigate rising emissions, so the construction of global help and co-operation and collective learning are needed to prepare for, and adapt to the unavoidable impacts.

Reduced consumption or a technological revolution?

An end to economic growth is a popular prescription among many environmentalists and it certainly has attractions. Some moves in this direction may contribute to a solution but are unlikely to be sufficiently effective on their own. Experience of the global financial crisis demonstrated to most people that a no growth economy, in a capitalist system, produces insecurity and unemployment. Reduced consumption by the populations of richer countries would need to be substantial and meanwhile global population growth is still coming through the pipeline and developing countries are still rapidly and visibly emulating the automobile and consumer culture of the richer countries.

As an optional choice by a minority, reduced consumption can make a contribution. Unfortunately, however, it is likely to be inadequate and unsustainable, with diminishing returns, if adopted freely by individuals, and to be corrupt, conflict ridden and authoritarian if imposed by force. Preaching to rapidly developing poor countries to cut their consumption is unjust, ineffective and offensive. Instead they should be helped to leapfrog the dirty technology of the Western Industrial Revolution via the sustainable technology that is still being developed.

Past exhortations to reduce consumption of alcohol, cigarettes or junk food had notoriously limited effectiveness. Sustaining voluntary or democratically chosen emission reduction initiatives, in the longer term, is made harder by the fact that those who do not participate or who break the rules are in effect free riders, who negate the achievements of others, making the whole exercise seem increasingly pointless.

Fashions for self-denial come and go. Periods of bacchanalia have indeed been followed by bonfires of the vanities, but in the end Savonarola was burned at the stake as the crowds cheered. Public moods can change rapidly, as those with vested interest in climate change denial foster popular scepticism. A local cold winter or drought-breaking rain can undermine confidence.

Most people may indeed find it easier to focus on the present, but it is doubtful if such short-sightedness is universal or inevitable. Fear of eternal damnation after death, or of the long-term sullying of a family's future reputation, have been powerful motivators throughout history. Over the last half century there has, however, been a cultural change, with a fore-shortening of time horizons for economic and political calculations and ever more rapid changes of careers, fashion and technological gizmos. It is hard to imagine the needed massive reductions in consumption being acceptable on a sufficient scale without tyrannical imposition, major social conflict and increasing inequality and corruption. Legal bans promote underground cultures, black markets and smuggling gangs.

The question of whether changed practices and renewable energy can be developed, on a scale to replace the greenhouse emissions currently being spewed into the atmosphere, is a matter of contention, rejected both by advocates of a no-growth economy and by many pessimists. In the following chapters we suggest that there are many technically viable or promising possibilities and that the obstacles are mainly economic and political. Their development might actually involve accelerated economic growth and activity, but also a diversion of resources from immediate consumption to long-term investment. Is human nature or a capitalist economy inherently incompatible with such a change?

Such a claim derives from an illegitimate amalgam and a historically truncated understanding. A short-term perspective is not an unavoidable product of either human nature or indeed of capitalism. Warriors have given their lives for glory after death. Cathedral construction that would take centuries has been undertaken in order to accumulate credit for eternal life. Self abnegation or migrating into the unknown in rickety boats, in order to give one's children the prospects for a better life, is understood by poor people around the world, and 'planting trees for one's grand-children' is a widespread ethical maxim. Long-term dynastic and empire-building motives have animated many political leaders, entrepreneurs and business tycoons.

Capitalism, growth and consumption

Some analysts in a green Marxist tradition argue that the capitalist system could not survive without continually increasing economic activity and emissions. It is, they say, addicted to growth and trapped by competitive market pressures on a treadmill of ever rising production of consumer goods. While some of them see this as necessitating an overturning and replacement of the system, others despair, seeing no currently available forces or models for such a transformation.

The bracketing together of the capitalist need for economic growth with a supposed need also for rising consumption is misleading. Recent capitalism has indeed been based on the production and sale of consumer goods, but this has not always been so. Periods of large-scale and profitable investment in capital equipment and infrastructure have seen significant restriction of consumption, as happened during the building of the railways. During the Second World War, the co-existence of full employment and reduced consumption was achieved through rationing and saving. After the war, the arms race and the space race provided profitable investment and employment opportunities, in industries whose products were neither sold nor consumed, thus avoiding a recurrence of the pre-war depression.

There is no necessary reason why a shift of resources from immediate consumption into research and development and a retooling of the energy and transport systems and the built environment, should be blocked by human nature or should endanger overall profits, economic growth or full employment, or portend the end of capitalism, although it will advantage some sectors and disadvantage others.

A more restricted question is whether this can be achieved by the kind of neoliberal free market capitalism that has evolved over recent decades. A pervasive aspect of the trends of the last thirty years, besides the elevation of the primacy of market logic, has been a further foreshortening of time in economic calculations, which increasingly favour short-term benefits and discount the future. Flexible response, facilitated by information technology, in a competitive and unpredictable environment, has replaced long-term planning. Short-term profits have become the main motivators, not only of stockholders but also for managers, whose bonuses have been linked to the next dividend returns. Computerised retail check-outs send instant messages about consumer demand. Product life cycles have shrunk and the time from conception to marketing has dropped dramatically, while new fashions and entertainment gizmos replace each other ever faster, all with their associated environmental effects.

The global economy is dominated by large corporations with increasingly short-term perspectives, in an environment where public regulations and policies have been substantially dismantled. The greatest handicap to a transformation driven by free market forces is, of course, the un-level playing field. Fossil fuel companies benefit from their established oligopolistic power and accumulated sunk capital, and their political influence has provided them with massive infrastructure and tax advantages at public expense. In the past they were able to suppress innovations such as gas fridges and the early electric cars. Above all they continue to benefit from their free externalities, where they pay nothing to compensate for their harmful effects on nature.

The business sector seems substantially divided or oscillating between those seeking new opportunities in innovation, and willing to accept government support and with it regulation, and those seeking to rest on their vested interests and accumulated advantages and to block new directions. Politics is also polarised by this issue. We shall return to this question in Chapter 11. At present many elements of a new energy economy are already appearing and competing for attention and investment, while attempts to level the playing fields for them are resisted by old vested interests.

Conclusion

Emission reductions need to start very quickly, for many reasons, but would also need to lead to large-scale, long-term goals. Climate change is already having negative impacts on people and eco-systems, and more is in the pipeline for the near future, demanding urgent mitigation. Above all we cannot be sure exactly how soon the accumulating pressures will reach irreversible tipping points. At the same time emissions remain in the atmosphere for centuries. An eventually uninhabitable Earth is not inconceivable. We must ultimately aim for a thorough going decarbonisation, a halt to all emissions, by a complete revolution in the way we produce and live.

Far more is spoken or written about mitigation than about adaptation. For those sceptical about climate change there is no need. For those hoping to forestall it, such talk would be a diversion. For those who have despaired, images of irrevocable disaster predominate. In general one of two alternative scenarios is offered. Either we will implement the needed swift and profound changes in the way we produce, travel and live; or else there is 'business as usual' and its ultimate consequences in physical catastrophes and irremediable social breakdown.

Given the deadlock in Copenhagen in December 2009, and the subsequent failure of measures to introduce a carbon price, through cap and trade measures in Australia and then the US, the second scenario is gaining credibility. However, those on the frontline of manifest climate change cannot indulge in such despair. They need to prepare immediately for the impacts, and learn to adapt and live with their consequences. Insofar as these are often poor countries with limited resources to protect themselves, their demands for assistance from those who are most responsible for the warming have been increasingly vocal in international gatherings. Their willingness to co-operate with mitigation plans such as halting deforestation and accepting lower-emissions paths of industrialisation may be strong bargaining chips. They have been led by the representatives of small island states, seeking help in building up their sea walls and obtaining promises of future havens for their populations if the inundation makes their lands uninhabitable.

In May 2010 Costa Rica's Christiana Figueres was chosen to be the new head of the UN climate convention, replacing outgoing chief Yvo de Boer. The implication of this choice are yet to unfold at the time of writing, but may be manifest at the next climate negotiations in Cancun, Mexico, in December 2010.

7

CHANGING OUR PRACTICES

Doing things differently

Efficiency

Efficiency measures are often promoted as low-hanging fruit, an easy and money-saving approach to reducing emissions that may be adopted rapidly by voluntary action, without impacting lifestyles, encouraged by promotion or subsidy or regulation. Many such measures are already in the pipeline. Rating systems help individuals to buy appliances and vehicles that consume less energy or fuel for the same effect; people are pressed to turn off lights and to disconnect computers and televisions after use; timers for showers and more efficient light bulbs are now readily available. A growing number of books, programmes and consultants advise people how to cut their energy consumption. Relying on individuals must, however, overcome the difficulty of initiating and sustaining changes in established habits and preferences. A British scheme sent 42 million energy-saving light bulbs free to homes. Although five times more efficient than the traditional incandescent ones, these were of unfamiliar appearance, and it is estimated most are lying in drawers unused.

A tightening up of efficiency standards can have an impact disproportionate to its costs. William Freudenburg argued, some years ago, that
generalising existing best practice between firms and nations could have
dramatic effects, as most of the harm was produced by a few bad actors.
Contrary to common assumptions, much environmental damage was not
economically 'necessary'. Data showed that, rather than producing
advanced materials, major polluters tended to be inefficient producers of
low-value commodities, and rather than being major employers, they
could have emissions-to-jobs ratios a thousand times worse than the
economy as a whole. The same disproportion could even be found
between nations with comparable levels of prosperity. In the 1980s the
United States consumed nearly twice as much oil, per unit of output, as
did West Germany (China consumed nearly three times as much).

China at present resists pressures to reduce absolutely its greenhouse
emissions, which could only be achieved by slowing economic growth.
It is, however, setting targets and making commitments to improve
its 'energy intensity', the amount of energy used per unit of output. Its
emphasis is on increasing efficiency in the 1,000 largest enterprises and on
retiring a wide range of inefficient industrial plants. China is also
improving fuel economy standards for passenger vehicle fleets, and its
standards are more stringent than those in Australia, the US or Canada
(although less stringent than those in Japan and Europe). The Chinese
government is also moving, at last, to extinguish the massive underground
coal fires in Inner Mongolia, which have been consuming an estimated 20 million tonnes of coal each year, for over fifty years. At a projected cost of US$100 million, this may be the most cost-effective
emissions reduction measure in the world.

Supermarkets provide a current example of a widespread high emissions practice with little or even negative usefulness. George Monbiot
reports that nearly two thirds of the expensive and environmentally
damaging electricity expended by the supermarket chain he studied in the
UK, was for refrigeration of goods, mostly caused by use of fridges and
even freezers without doors or lids, and by the need for extra heating to
counteract the discomfort they produced. The sole motive for the
absence of doors was to encourage impulse buying.

There are some promising new ways of increasing the efficiency, and
therefore reducing the amount, of fuel used. Cargo ships can reduce fuel

use by 30 per cent by travelling more slowly, and when oil prices are high many find this an economic choice. 'Chameleon' paint, for the outside of buildings, can be switched from a heat-absorbing dark colour in winter to a reflective light colour in summer. Another example is the Bloom Energy and Ceramic Fuel Cell, launched in 2010 for small-scale energy generation in homes and offices. This is claiming to convert 50–60 per cent of a wide range of fuels into electricity, compared with 28 per cent conversion of the energy in coal fired power stations. Japan's Toray Industries is supplying Boeing with lightweight carbon fibre, to replace steel and cut fuel consumption in its Dreamliner aeroplanes. Improving the future efficiency of new energy-generating technologies will be crucial to their competitiveness and widespread adoption.

Strict rules, mandating high insulation standards for all new buildings, are obviously essential. In developing Asia urbanisation is adding 44 million people to cities each year and every day sees the construction of 20,000 new dwellings. In already urbanised societies such rules will take a long time to have a wide effect. At current rates of construction, it is estimated that 85 per cent of existing buildings in Britain will still be standing in 2050, although the impact will accelerate as old structures are replaced. Retrofitting home insulation is therefore sometimes advocated. German regulations set energy efficiency standards that must be met, with the help of grants and cheap loans, as a condition for permission to renovate an old house, and 800,000 homes have been retrofitted since 2006. President Obama's stimulus package in 2009 set aside $5 billion to insulate 500,000 of America's poorest homes within two years.

However, large-scale retrofitting programmes are costly and require careful regulation and approval procedures. A disastrous result from an over-hasty scheme was the Australian government's stimulus measure to encourage job creation by retrofitting home insulation. The subsidy was very generous and a million applications were made over just three years. The scheme literally exploded in the government's face. Incompetent and even fraudulent installers, with inadequately trained workers, using unsafe materials, sprang up. There were four deaths by electrocution of young workers installing aluminium foil in roof cavities, and at least 120 house fires due to the use of defective inflammable Pink Batts. The programme was terminated, as the discredited minister explained that his department lacked the personnel and expertise to supervise it properly. The government had to offer free inspections to 150,000 householders and compensation to the respectable parts of the newly growing industry and its suddenly redundant workers, when the scheme was hastily wound up. The UK is now backtracking on its costly plans for energy-efficient homes.

An emphasis on efficiency is important in reducing emissions in the short run, but is not necessarily easy or cheap and will tend to have diminishing returns and a limited scope in the long term. It should not pre-empt the early and perhaps initially more expensive steps to more fundamental transformations of our energy systems.

Transport

Making public transport attractive and reliable is an important first step. Very fast bullet trains are well established in Japan and Europe, and China will soon have more high speed track than the rest of the world combined. In France the SNCF has plans to rebuild its 3,000 stations as attractive multi-use urban and retail hubs, using bio-climactic systems and new, recyclable materials, so they can sustain comfortable temperatures in summer and winter, without consuming energy.

The Obama administration is now launching plans for high speed rail, which has so far been undeveloped in the US. The plans envision a national network of fast inter-city corridors, offering an alternative to long car journeys and short flights. The first $8 billion was inserted into last year's stimulus package, although this is still only one eighth of last year's federal spending on highways. European train makers are showing interest in the new markets this will offer them. It is, however, a small instalment on the cost of a new fast national network that some have estimated as high as a trillion dollars.

Certainly, better public transport and more walking and cycling can have immediate effects, but limits are placed on how much this can achieve, in many developed societies, by the physical topography of suburbanised cities, an ageing population and a car-focussed consumer culture. In successfully developing societies the longing to emulate the automobile culture of developed countries may be irresistible, at least in the short to medium term, and reducing car emissions is probably the first thing that comes to most people's minds when thinking about mitigation of greenhouse gases.

Hybrid electric cars are already available, and all major car companies are planning to put all-electric car models on the market by 2011. At present they are relatively expensive and their use is restricted by a dearth of recharging outlets. Lithium batteries will increase their efficiency and bring down the price. Shai Agassi's Better Place is working towards providing a network of battery swapping stations with operations projected in Israel, China, Japan, Canada, the United States, France, Denmark and Australia. They will help curb emissions, but until the electricity that drives them comes from non-emitting sources their long-term impact will be much reduced. Better Place promises to recharge its batteries only from carbon free sources, but such clean energy generation will have to be enormously expanded if it is to provide a general replacement for petrol as well as coal. So far no substitute for diesel for aircraft has been developed, although some algae-based biofuels may be viable at low temperatures. Fast trains can replace many flights. Emissions from world shipping are substantial, but there is talk of a partial return to wind power, using giant kites when appropriate.

Changing agricultural practices

A United Nations report says 18 per cent of greenhouse emissions are caused by methane in livestock flatulence. A cow is responsible for fifteen times more methane than a sheep, and changed eating habits could help, particularly to kangaroo meat. The Australian and New Zealand governments are also investing in research into improved feeding and breeding to reduce the methane belched.

Rice, the most heavily consumed staple food on earth, is a major emitter of the greenhouse gas methane. In the year 2000 it was responsible for some 625 million metric tons of carbon dioxide equivalent. The International Food Policy Research Institute (IFPRI) estimates such emissions could be substantially reduced by different farming practices such as changing from prolonged flooding of fields to intermittent irrigation and mid-season drainage or, where possible, to rain-fed rice, and adding organic inputs while the fields are not flooded.

Another potential is agro-forestry which combines woody perennials with herbaceous crops and/or animals, either in some form of spatial arrangement or temporal sequence on the same land. IFPRI estimates that it can sequester five times more carbon from the atmosphere than ordinary croplands. A major greenhouse gas is the nitrous oxide emitted by nitrogen fertiliser and by bacteria in acidic oceans. Finding substitutes that would not damage food production is a huge challenge. One possibility requiring more research is biochar (agro-forestry and biochar are discussed further in Chapter 10). Other research is seeking to grow crops with light-coloured leaves that reflect the sun.

Reducing deforestation

Deforestation, like ocean acidification, limits the capacity of natural sinks to absorb our carbon, as well as contributing directly to greenhouse gases, through emissions from the decay of plants that are not replaced by new growth. It is credited by the IPCC with 17 per cent of the increase in global emissions.

In what was billed as 'the biggest commercial forest-preservation agreement in history', environmental advocates and timber companies, in May 2010, struck a deal to protect two thirds of Canada's forests, 72 million hectares, an area twice the size of Germany. The pact will ban logging in parts of the area and limit it in others to 'world best practices' would be implemented. The pact covers a bigger area than the previous global leader, the Brazilian Amazon Region Protected Areas project, where satellite observation is used to identify illegal logging, reducing deforestation by one-third over five years.

Deforestation was not included in the earlier global agreements under the Kyoto protocol, but is currently moving to the top of the agenda. In May 2010, rich and poor countries at the Oslo Climate and Forest Conference, agreed to guidelines for releasing aid for Reducing Emissions from Deforestation and Degradation (REDD) programmes in developing countries. This was the first concrete sign of global action since the Copenhagen impasse the previous December. Both developed country governments and private investors, seeking carbon credits, would contribute. They agreed to set up monitoring mechanisms to ensure that illegal logging was controlled, that there were real reductions in what would indeed otherwise have been logged and that funds really went to local farming and forest communities. This is a tall order yet to be verified, and some have accused the scheme of being backed by 'big polluters and climate profiteers'. In a first step Norway announced a billion dollars for Indonesia, and President Yudhoyono announced a two year moratorium on converting peat and forest land to agriculture.

Carbon capture and storage

The idea of capturing the CO_2 emitted by burning coal, compressing it and burying it safely underground or at the bottom of the ocean, has appealed at various times to governments in the USA, Australia, Europe and Asia. Coal is the cheapest fuel, has driven industrialisation throughout the modern era and remains central to the growth plans of developing economies. In countries with significant reserves, it is a major employer, guarantor of energy independence and source of export earnings, and still has enormous political clout. But is this solution to greenhouse emissions too good to be true? Why is there still no big power plant using Carbon Capture and Storage (CCS) anywhere in the world?

There have been public and media fears about the feasibility of such capture, about the availability of sequestration sites and about the dangers of leakage leading to poisoning, explosions or negation of the benefits. Independent energy technologists and geologists, however, are generally confident that many such projects are do-able. The capture and injection process has already been demonstrated. Oil companies already use most of the processes involved, pumping CO_2 underground in order to pressure the extraction of remnant oil and gas. Humans have been hollowing out coal, oil, gas and minerals for centuries, leaving spaces in the earth and the ocean floor, and there are many other sites with suitably porous rocks. Oil, gas and salt water seem to stay put in certain rock formations indefinitely. A study, reported in *Nature*, of carbon gas naturally buried, long ago, in nine natural gas fields in America, Europe and China, found they had not leaked in thousands or millions of years.

There are still technical problems to be resolved, particularly about what would be involved in a really large-scale operation, with very many diverse sequestration sites These are not, however, enough to explain why, despite the political hype and subsidies, so little progress has been made, with so many projects stalling and so little private investment flowing in. An audit of efforts to demonstrate feasibility, carried out by the Australian Carbon Capture and Storage Institute in 2009, found 275 different projects around the world, but only thirty-four completed and only seven commercial-scale projects in operation.

ZeroGen in Australia, hyped as a world leader, announced a successful trial in 2007 and projected a working plant by 2011. In 2010 this projected start was delayed until 2017. FutureGen, the only large Carbon Capture and Storage demonstration project planned for the US, was started in 2003, but five years later the Department of Energy cancelled its commitment, because of delays and spiralling costs. A journalist visiting their proposed site reported an empty field. Its funding has now been restored in Obama's 2010 budget.

The impact of the technology may even be negative, providing an excuse to allow new coal burning power stations under the pretence of their readiness to be retrofitted for carbon capture at some hypothetical later date. Once constructed their sunk costs would guarantee their continuation for many years. In Britain the Electricity Act requires that from November 2009, applications for construction or upgrading of coal-fired power stations must provide evidence that the plant will be capable of demonstrating CCS on at least 300MWe of its capacity from the outset. Since a typical plant has a capacity of 1,600MWe this seems more like a get-out-of-jail-free card than a serious mandating of CCS.

The problems seem to be mainly economic – the costs of production are higher than for conventional coal firing, or for competitors in the production of electricity such as oil or gas, or even for solar or wind power as their prices come down. Some of this would be reduced if the schemes took off on a large scale, but some are inherent to the needs of the process, which consumes some 30 per cent more coal for the same electricity output as current methods. A price on carbon (through a carbon tax or emissions trading scheme) is argued by many to be essential, but would need to be much steeper than even the current European price if it were to create a level playing field with uncaptured coal-fired power. Other inhibitors of the private sector investing in such long-term and large-scale projects are the shifting uncertainty of public subsidy and the fluctuation in the price of competitors such as oil.

Is it worth it? Carbon capture is currently making no contribution to mitigating climate change and is unlikely to contribute much for at least several decades. What about the long term? Could it be the mechanism for constructing a decarbonised but prosperous world economy? Here we must raise the question of the extent of world coal reserves. There is certainly enough coal to tip the world into climate catastrophe if all of it is burned as now, but if the CO_2 is captured, can the reserves guarantee safe energy for a prosperous decarbonised human future?

World coal reserves in 2006 were estimated at around 900,000 million tons. At current extraction rates this would last for 130 years, but allowing for the same kind of demand growth as we have been seeing, particularly if electric cars become widespread, they would not last much beyond the middle of this century. The additional coal consumption of the capture and storage process would add yet another 30 per cent to these depletion rates.

Some new finds are of course probable, but coal has been sought for so long that it is unlikely these will be very large or easy to extract. Just as with 'peak oil', when the most accessible and high quality seams are exhausted prices would rise, supplies become more erratic and unreliable and import dependence would grow. A visiting Chinese scientist explained to the author that they had limited interest in major long-term investment in researching or buying carbon sequestration technology because they were approaching the limits of their coal reserves. They were already importing coal from Australia, and concerned about the current atmospheric pollution, mine accidents and future security of supplies, and looking rather to develop nuclear and renewable energy sources.

Carbon capture has disappointed expectations in the short term. In the long term its costs, and the prospects of rising coal prices and shrinking reserves, will be a major disincentive to major investment in many countries, in particular those whose domestic supplies are depleting. A few countries, however, in particular the United States, South Africa and Australia, have large reserves relative to the world average and to the size of their domestic economies, and highly developed and powerful coal mining and generating industries. Here carbon capture for domestic energy production may have some viable prospects in the medium to longer term. In these cases there should be strict regulation to ensure that any refurbishing or new construction of coal fired power plants involves complete carbon capture and storage, from the start of their functioning.

The most pertinent fear, expressed by many opponents, is that sequestration projects will in the end prove impractical or uneconomic, but meanwhile their pursuit will spread complacency and green-washing, allow more coal burning, and divert public funds away from more viable remedies.

8

CHANGING OUR POWER I

Natural gas, biofuels and nuclear energy

Economic and political rather than technical obstacles are the main hurdles obstructing the take-off of a low emissions energy revolution. Which forms can be developed to be socially acceptable and competitive with the powerful fossil fuel interests, is yet to be demonstrated, and the answer is likely to be different in different parts of the world. Some of the options will be discussed in this and the next chapter.

Natural gas

'Conventional' natural gas

An important if limited mitigation has been obtained by conversion of cars to run on liquid petroleum gas, which emits 15 per cent less greenhouse gas than petrol, and by conversion of power stations to natural gas, which emits only a third as much as brown coal and half as much as black coal, with even lower proportions of other pollutants. At the least any new or renovated power stations could, from the start, be planned to burn gas not coal.

The gas has often been found near oil fields and coal beds, can be transported in conventional pipelines and tankers, and requires no major modifications to existing coal-fired power stations. The largest reserves in the world are in Russia, the Middle East and Africa, and conversion from the use of coal to gas may conflict with a European desire to promote energy independence, which unfortunately might favour domestic coal. In the UK natural gas has grown from just 0.5 per cent of the electricity generation mix in 1990, to 45 per cent in 2008, and carbon emissions associated with the generation have fallen there by 26 per cent. Australia has signed binding trade deals to supply India with $25 billion worth of gas, and with China to supply 3.6 million tonnes per annum for twenty years, starting in 2014, worth $60 billion. Australian exports of coal far exceed this, and plans for renovating dirty domestic power plants still focus on coal.

However, gas is often wasted, due to unwillingness of oil companies to invest in the necessary infrastructure. Particularly harmful is the practice of flaring the gas, which can severely damage the health of local populations and pours greenhouse emissions into the atmosphere for no useful purpose. In Nigeria it is estimated the gas thus flared is equal to a quarter of Britain's power needs. Known global reserves at current levels of use would run out by 2068, and much sooner if it was more widely used to replace coal.

Shale gas

In June 2010, Helen Knight published an article in *New Scientist* describing recent major breakthroughs in the extraction of natural gas from shale, using a horizontal fracturing ('fracing') technique. This involves pumping in millions of litres of water to make the shale permeable, but much of this can be recovered and reused. Shale gas would double known global reserves and could provide many parts of the world with energy independence. It is claimed it might be possible to reduce the emissions from the gas further, by capturing and storing the carbon dioxide before the gas is burnt, although to date the focus of carbon capture and storage research has been on coal. The viability of these new sources will depend on finding ways to minimise water usage and ground water contamination. If so, Knight writes, 'natural gas will be providing us with clean energy for many decades to come'.

Biofuels

Petroleum fuels are responsible for over a third of CO_2 emissions and for a time biofuels were seen as an important source of mitigation that, unlike fossil fuels, would re-absorb their own emissions into the following year's crop. It was hoped they could provide at least a short-term transition before decarbonised electricity generation could charge a new generation of electric vehicles. The conversion of trains to biodiesel could cut their emissions by three quarters, at little extra cost.

Biofuels are argued to be the only carbon neutral alternatives to petroleum and diesel fuels. They expanded rapidly in recent years, subsidised or mandated by governments who wished to provide new markets for their farmers and to overcome dependence on imported oil. In 2008 Brazil produced nearly 25 billion litres of ethanol and Europe produced nearly 12 billion litres, and biofuels supplied 3.5 per cent of US consumption. New vehicle models, such as the Toyota Prius and the Holden Volt, that could use biofuel blends, have been launched successfully.

These first generation (Gen-1) biofuels were either made from food crops or grown on agricultural land, and included ethanol from cereals and sugars and diesel from beans, seeds and vegetable oils. In 2008 world food prices spiked sharply. This may have been a boon for many poor as well as wealthier farmers, but it had a severe impact on the urban poor around the world, as well as on the capacity of aid organisations to provide for the displaced and those affected by droughts and disasters. Protests and riots followed.

The rapid rise in global food prices was attributed to the diversion of crops and land to biofuels, and while this certainly played a part, other factors did also contribute. The pouring of billions into speculative futures markets led directly to massive stockpiling of key commodities, such as corn, whose price increased tenfold between 2003 and 2008. The rising cost of oil also had a severe impact on the price of agricultural fertilisers and transport. Instability in the global oil price made long-term planning and investment in biofuels difficult. Competition with rising demand for food raised the costs of biofuel feed stocks, but a subsequent fall back in the price of oil only increased competition from petroleum for their final product.

As the reputation and the economic viability of the industry suffered, its future growth was seriously compromised. In Australia the industry collapsed in the financial crisis in 2007 and 2008 and received no rescue from banks or government. Yet there may be other more substantial long-term prospects in the development of Gen-2 biofuels that do not compete with food production.

The initial rapid growth of the Gen-1 biofuels industry proved very easy to set in motion but was highly problematic in its competition with food for crops and scarce agricultural land. It has been a short-term boon to farmers but, with a growing world population, might at most replace only 5–10 per cent of current petroleum fuel. There is not enough agricultural land in the world to replace all the world's future fuel requirements.

Researchers Stephen Clarke, Dan Graiver and Sudirman Habibie argue that second generation (Gen-2) fuel extraction, with feedstock from closed micro algae farms in marine, brackish or waste water, from jatropha plantations in desert or marginal land, from waste and from hydrolysis of ligno-cellulosics such as straw, bagasse and woody biomass, may have much more long-term potential, although technical and cost problems remain to be resolved. Some initial experiments suggest that even fuel suitable for use by jets, in the cold atmosphere of high altitudes, might also be possible.

Micro algae are thirty times more efficient converters of solar energy than terrestrial feedstock. Indeed most of the world's carbon dioxide is currently sequestered by algae in oceans. However, such farms are not currently economically competitive with petroleum. This is mainly because they are still at an early stage of development and because of the costs of establishing remote and scattered sources and of transporting the product to end users. They argue that the development of bio-refinery co-products such as animal feed and fertiliser, and especially of valuable chemicals from the same algae-based feed stock, could underpin the cost of this fuel, but more work is needed on this. Much marginal land around the world (including for example as much as 14 million hectares in Indonesia) could be suitable for jatropha curcas plantations. It is estimated the plants could provide 1.5 tonnes of raw oil per hectare and have a productive life of up to fifty years.

Short-term economic calculations favoured Gen-1 biofuels, with disastrous results, while the likely long-term nature of benefits of Gen-2 sources has found it hard to attract research funding and investment. The fall in the price of oil during the global financial crisis also discouraged biofuels investment. However, there are some signs of a turnaround with the recent establishment of a Biofuels Institute in Britain and three new major research institutes in bio-energy in the US, and with development projects by companies such as Coskata and Dow Chemicals and investments by GM and even ExxonMobil.

Nuclear power – decline and rebirth

Although long established, nuclear power made only a marginal contribution to electricity generation at the start of the twenty-first century, except in France and Sweden. Legislation to ban its expansion or to phase it out had been passed in Sweden and also in Italy, Belgium, Austria, the Netherlands and Spain. By the end of the first decade no new nuclear power stations had been completed in Europe since 1991. The largest nuclear energy company in the world today is state-owned Areva in France, which operates fifty-nine reactors and supplies 80 per cent of the country's electricity.

In the US there were still 104 ageing reactors operating, but private investment had halted in the 1970s, and 138 planned reactors were cancelled in the 1970s and 1980s. No new plants were started after the Three Mile Island disaster in 1979. Training, research and development were cut back, and expertise has atrophied. This was due to a convergence of economic and political reasons. Costs remained high, despite huge direct and indirect government support and subsidies, estimated to be worth over $500 billion, equal to $15 per kw generated between 1947 and 1999. By the mid-1980s, competing oil prices were falling back again after the oil embargo spike and the Islamic revolution in Iran, and growth in demand for energy slowed in response to recession, conservation and efficiency policies. The nuclear handicap was exacerbated by the uncertainties and long delays in approval, caused by public fears in the US, as elsewhere, of radiation from leaks, nuclear waste and accidents.

By mid-2010, however, the number of reactors under construction globally had risen to fifty-nine, up from forty a year earlier, with twenty-three of them in China, eleven in Russia, six in South Korea and four in India. The number of functioning reactors was expected to grow from

425 in 29 countries, to 568 in 42 countries by 2020. Italy has just reversed its ban on construction and Germany has postponed phasing out ageing plants. Sweden had voted to phase out nuclear power in a 1980 referendum, but only two of the country's twelve reactors have been closed. The government says new reactors are needed to help fight climate change and secure energy supply. Ten reactors still supply about 50 per cent of the country's electricity, and in June they decided gradually to replace them. Public support for nuclear energy has grown there amid concerns over global warming and the reliability of foreign energy suppliers. The UK in November 2009 approved ten new sites, and promised to fast track permissions. In the US, in February 2010, Obama announced federal loan guarantees for two reactors to be built at an existing nuclear facility in Georgia, the first for thirty years. In March Russia offered nineteen reactors to India.

Advocates argue that nuclear power generation is a mature technological alternative to fossil fuels, already in widespread use and ready to go. It is claimed, perhaps rightly, that disasters like that at Chernobyl could not happen with the current safety designs, and that its health and safety risks are less than those caused by coal-based energy.

It can be argued that the world's need for immediate climate change mitigation can be assisted in places such as China by substituting nuclear power for what would otherwise be new coal burning power generators. China is the largest consumer of coal in the world and derives over two thirds of its power from coal generation. This is perhaps the most important single driver of rising global greenhouse emissions. Given the size of the country and the speed of its economic growth, it is improbable that renewable energy could provide a significant substitute for coal in the short term. Three of the fifty-odd nuclear reactors under construction in China are Gen-3 reactors. These, being built by Westinghouse and Areva, will be the first completed of their kind in the world and are claimed to reduce substantially the risks, waste and inefficiency associated with Gen-2 models. For good or ill, useful lessons may be learned from the experience.

There are, however, serious doubts about the potential of nuclear power, as currently developed, to provide long-term and widespread replacement of fossil fuels. The long-term benefits of current nuclear power technology are likely to be limited. Known reserves of uranium could be rapidly exhausted if existing forms of nuclear power became a widespread substitute for fossil fuels. In 2005 it was estimated that existing known sources of high grade uranium would only last a decade or so if they were used to replace all coal fired power stations.

There are claims by scientists of continuing health risks to those living nearby. The British government's Health and Safety Executive, in November 2009, refused to support the plans for new reactors. Problems of transporting and storing or reprocessing waste, that can remain radio-active for hundreds of thousands of years, have not been resolved. The only US-proposed site for long-term storage, at Yucca Mountain in Nevada, has been cancelled because of fears of geological instability, leaving existing waste stockpiles still awaiting a long-term home. In Germany, nuclear protests against extending the lifespan of existing nuclear power stations have been spurred by revelations of the possible collapse of an old waste dump in a leaky salt mine in Saxony. France relies on controversial and expensive waste reprocessing. Weaponization by maverick states or terrorists may be made easier by advances in enrichment techniques, a danger which could justify a massive expansion of state control.

Such doubts feed public opposition, causing delays and escalating costs. It is by no means clear that the very substantial expenditure required to develop nuclear power on a large scale would not be more productive, in terms of cutting emissions in the long term, if invested in other forms of existing alternative and renewable energy or in a large expansion of research into long-term alternatives.

Current nuclear power technology may help in the short run, but if dangers of accidents, waste or weapons proliferation mount, or world uranium supplies are exhausted, this might prove not only dangerous but also a dead end for any further expansion.

Nuclear's long-term promises?

The initial promise of nuclear power was of unlimited, cheap, clean and non-polluting energy. Existing nuclear generation technology poses many short-term problems and could not be scaled up to a major role in long-term decarbonisation. It is, however, possible that as yet untried break-throughs in research on Gen-4 fast reactors, thorium-based reactors or nuclear fusion, could reinstate the promise, in forms that would not contribute to radioactive waste or weapons proliferation and which would use less uranium or none at all. Many dismiss the possibility as decades away from practical application (if ever), yet we negotiate emission reduction targets up to 2050, and our need is for a long-term transformation as well as immediate measures.

The stagnation and decline of nuclear energy generation, in the 1980s and 1990s, left much research and development stranded. In the US the number of academic nuclear engineering programmes fell by more than half, and work on new approaches lost momentum. In recent years there has been a revival, and a strong case can be made for significant expenditure on such research now.

The most work is being devoted to what are called Gen-4 reactors, with research initiated by the GIF (Generation 4 International Forum). The GIF is an international collective, formed in 2000, representing governments of thirteen countries where nuclear energy is seen as significant now and in the future, including the US, the European Union, Russia and China. The goals have been to improve safety, reduce waste, resist proliferation and cut costs. They have chosen to work on a set of six new reactor designs with the first intended to be available for commercial construction by 2021. Relative to existing Gen-1–3 reactors, their promised benefits are to leave nuclear waste that lasts only decades instead of millennia; 100–300 times more energy from the same amount of nuclear fuel; and the ability to consume existing stockpiles of nuclear waste.

The use of thorium, rather than uranium, as a nuclear fuel was studied for some thirty years on a small scale and fuelled some early nuclear reactors in America and Russia. R& D was discontinued in the US in 1973, when the US nuclear industry decided to build nuclear plants using uranium. The plutonium by-product was wanted by the military for nuclear weapons, for which thorium was not suitable. Today thorium fuel cycle research is taking place in India, in particular, and also in Norway, Canada and Russia; and its revival has active and vocal supporters in the US. Thorium is an abundant, widely and easily available naturally occurring fertile material, the only other one on earth beside natural uranium. It is found in small amounts in most rocks and soils. The World Nuclear Association sees its main advantages in its abundance and its very limited and short lived waste. There are high costs of fuel fabrication but it is resistant to proliferation. Much development work is still required and some technical problems remain, but it is said to hold considerable potential in the long term. Advocates claim it is incapable of causing a meltdown; does not require expensive conversion or enrichment; has waste that can mainly be recycled as fuel, with the remainder radiotoxic for only tens of years; and that it can be used in existing nuclear reactors.

Fusion energy, a process that mirrors the process taking place within the sun, has been a notorious mirage, always 'just a few years away' for decades. It biggest problem has been how to produce more energy than is used in the process. A recent article in *Science* claimed a major hurdle to the use of lasers to initiate fusion had been overcome at the recently opened National Ignition Facility, at the Lawrence Livermore National Laboratory. Experiments aiming at fusion were scheduled to start before May 2010, with the claim by the team leader that 'it is going to happen this year'. In October the National Ignition Facility declared successfully completed an integrated ignition experiment, an important step towards fusion ignition. Many remain sceptical. Meanwhile the French Iter fusion project is facing escalating costs and delays, with its demonstration project now postponed from 2015 to 2019, the first fusion reaction not expected until 2026, and electricity generation not expected until 2040.

An article in *New Scientist* in March 2010, by Julian Hunt and Graham O'Connor, 'Hybrid fusion: the third nuclear option', argues that 'fission is unsafe and fusion is decades away, but put them together and the problems [including that of uranium scarcity, long-lived waste and weapons proliferation] melt away'. China's fusion research centre is planning to build a prototype of the hybrid by 2020 and interest is being shown in Europe and America.

9

CHANGING OUR POWER II

Water, wind, sun and earth

Power from flowing water, blowing winds and shining sun and from the heat buried deep within the earth are in principle effectively inexhaustible and carbon free. A study published by the US National Academy of Sciences in 2009, calculated that a network of land-based 2.5 megawatt wind turbines, restricted to non-forested, ice-free, non-urban areas, operating at as little as 20 per cent capacity, could supply more than forty times the world's current electricity use and more than five times the total global energy use. It is estimated that solar energy could supply all of Australia's needs using just 50 square kilometres of desert. A 250km-wide patch of the Sahara could meet all of Europe's electricity consumption. A major study by the Massachusetts Institute of Technology estimated that the 'technically extractable portion' of the US geothermal resource is 'about 2,000 times its annual consumption of primary energy'. The total global geothermal resource is some 280,000 times the world's annual consumption of primary energy.

Some of the technology to tap these sources, such as hydro dams and windmills, is very old but can be much further developed. Some, such as extracting power from tidal, wave and solar energy and from geothermal heat, is still expensive and inefficient and in need of new economies of scale and of reliable long-term investment in infrastructure, research and development.

Renewable energy

Hydro-electric power

Hydro-electric projects can be massive, with unpredictable environmental consequences. Environmentalists are deeply split on whether they should be resisted, as currently in Brazil, or accepted as a necessary evil. Attempts to dam the Franklin River in Tasmania were blocked by environmental protests in the 1980s, launching the Australian green movement.

Power generation by hydro-electric dams has a long history in select places and provides 18 per cent of the world's electricity and most of the domestic electricity needs of Norway, Brazil, Venezuela and Paraguay. Africa is the most impoverished continent, and its 800 million people are significantly starved of power and the development it can bring. Its potential for hydro-electricity is substantial, perhaps left dormant because, unlike Africa's other resources, it cannot just be ripped out and transported far away. Technological development and the rising price of oil have now put the construction of such dams on the agenda. Ethiopia has plans for many projects, with the largest being to dam the source of the Blue Nile at Gibe III. It hopes to multiply power generation fifteenfold by 2020, meeting all its own needs and becoming a significant exporter to neighbouring countries. The Congo is also proposing massive projects. How feasible any of these are in a continent of failed states, wracked by conflict, is yet to be seen but perhaps they may offer a circuit-breaker.

The Himalayan glaciers feed the rivers that already generate much of Asia's hydro-electricity. China obtains 22 per cent of its electricity from hydro power and most of its renewable energy targets are to be met by many new hydro-electric projects. The Three Gorges Dam in China is the world's largest such project, costing US$30 billion, aiming to generate electricity and control floods. It was started in 1993 and completed in 2008. Its total generating capacity will eventually reach 22,500MW, saving 100 million tons of CO_2 emissions each year, compared to coal generation. More than a million people were relocated for its construction. There have been warnings of severe ecological damage and of a possible need to move 4 million more people because of the danger of serious landslides on the steep hills around the dam. Recent climate change effects have reduced the flow of water, interfering with the electricity generation on many smaller dams.

The potential for hydro-electricity to contribute further to carbon-free energy is argued in new dam proposals on the Mekong and in PNG. On the one hand they can be a major long-term source of renewable energy. On the other, negative impacts include temporary high emissions from the huge quantities of cement used in their construction and from the rotting flooded vegetation, as well as other negative social and environmental disruptions. If glaciers dry up, hydro power will prove a short-term boon and a long-term dead-end.

Tidal and wave power have been considered marginal possibilities, but are now becoming more prominent, with South Korea in the lead. Their Sihwa tidal barrage, with an output of 254MW, converting an existing sea wall, is due for completion at the end of 2010, and there is a much larger proposal for Incheon, due for completion in 2017. The Netherlands are also investigating converting their existing dykes for tidal generation.

Wind power

This has been the most recently successful of the new generation of renewables. Energy experts said the sector had maintained a near 30 per cent annual growth rate globally in 2008 and they expected the industry to employ 1 million people by the end of the decade. Germany, the global leader, had focussed on co-operation and profit sharing with local communities, to establish its lead in land-based wind turbines, but lost ground to Britain's new focus on offshore turbines. Meanwhile new, low maintenance, lightweight giant turbines, specially designed for installation in the sea, have been developed for Germany's first offshore wind farm, putting them back in the game.

The full potential of wind power has yet to be realised. At the end of 2008 Africa's installed wind power capacity was less than 600MW. However, in Kenya, a Dutch consortium has plans to install 365 giant turbines in the desert around Lake Turkana, creating the biggest wind farm in Africa, whose capacity of 300MW, when completed in 2012, will be equal to 25 per cent of Kenya's current installed power from all sources. Tanzania has plans to generate 100MW of power and Egypt has declared its intention to supply 12 per cent of its energy needs by wind power by 2020. Even coal-intensive South Africa is offering consumers a feed in tariff for wind power. The realisation of these plans has yet many hurdles to overcome.

Solar power

Sunlight on roofs can directly and effectively provide domestic hot water, although it needs some backup from mains electricity in overcast weather. This was widely adopted in many parts of suburban Australia during the oil price spike in the 1970s. The sun can also generate electricity in two main ways. Solar cells in panels on rooftops convert sunlight directly to power, but at a low level of efficiency. They can be mass produced and then sold to small users and households. Although at present more expensive than the second method, it has more long-term innovation potential and its costs are estimated to fall by 20 per cent for each doubling of production. The panels have now been found to last at least thirty years, and the cost of obtaining electricity in this way is predicted to reach parity with grid electricity for half of Europe by 2020.

The second method is concentrated solar thermal (CST) which involves use of mirrors, generally in desert areas, that focus the sun's heat onto a central, steam-driven turbine. This is at present the cheaper, simpler and more efficient of the two technologies but it requires a large area and strong sunlight, and has less potential for technical improvement. Nine early CST plants, using 2 million square metres of mirrors, were built between 1984 and 1991 in the Mojave Desert of Southern California. Backed up by gas-fired generators for sunless times, they have operated successfully for a quarter of a century. In the first decade of the twenty-first century new and improved CST plants have been built in America's west, and many more are under construction, including a utility-scale photo-voltaic plant in Florida. Nevada's Solar One power plant, dedicated in 2008, covers 400 acres and powers 14,000 homes. In Germany, Spain and Italy government incentives have given a greater than predicted boost to solar energy, and China and Taiwan are seeking to become world leaders in production of photo-voltaics. The Sahara, the 2.5 million km^2 Tibetan plateau and much of central Australia are huge, lightly populated and barely cultivated areas, on which the sun beats down for significant parts of the year.

Nonetheless there have been some setbacks. In 2008, half of all solar power installed globally was in Spain. The highest subsidies anywhere attracted companies from around the world and fuelled a brief boom in often low quality, poorly designed solar plants, likely to require indefinite support. In 2010 the government reacted by cutting payments and capping construction, and boom turned to bust. However, the most robust companies survived the downturn, restructured and are re-emerging as profitable enterprises with growing exports. In Australia after a period of promising growth, some major solar manufacturers moved to other countries, and it was revealed that European solar companies were abandoning possible solar projects in Australia because of a lack of consistent and supportive government policy.

China is aiming to lead the world in solar energy production. In 2009 the country invested about US$34 billion in solar panels, wind turbines and other alternative energy technologies, nearly twice as much as the US, although its relatively modest green gains could not keep up with the continuing rapid expansion of its coal- and oil-fuelled economy.

Geothermal power

Al Gore writes with great enthusiasm about geothermal power, potentially the largest source of energy in the world. The earth's hot zones, where tectonic plates meet, are vast in the 'ring of fire' surrounding the Pacific Ocean, across the Mediterranean and the Middle East and down the east coast of Africa, and there are numerous other scattered hot spots. Quite a lot of it is accessible using current mature drilling methods, developed for the oil industry, but many deeper sites are also becoming available as the technology progresses. The cost of energy, from sites between 3 and 6km down, is said to be competitive with other sources. Unlike wind and solar it is not intermittent and can provide base load. Research was halted by the Bush administration but is now resuming. Reservoir testing at a commercial scale remains to be done before private investors in the US will accept the large financial risks.

Meanwhile projects are under way in Europe – in France, Germany, Switzerland, the UK and the Czech Republic – and in Australia, where seven publicly traded companies are active. The Philippines, El Salvador, Costa Rica, Kenya and Iceland have recently achieved production of more than 15 per cent of their electricity, and New Zealand, Indonesia, Nicaragua and Guadaloupe have achieved between 5 and 10 per cent from geothermal sources. Indonesia hosted 'the world's biggest geothermal energy conference' in April 2010 and is hoping to raise more than US$1 billion in investment to further develop its geothermal energy.

Problems and solutions

Problems

Many problems have so far inhibited the widespread substitution of renewable energy for fossil fuels that many had hoped for. Their costs are still generally higher than fossil fuel competitors, who pay little or nothing for the environmental damage they cause. Like many innovators they cannot become economically competitive until they operate on a large scale, but they cannot achieve such scale until they are cost-competitive. They thus depend on initial government support, but this can be unpredictable, as can the see-sawing prices of competitors, particularly oil. Arguably a real breakthrough awaits a significant and stable global price imposed on carbon.

These sources also often suffer from intermittent energy and inconvenient locations. The sun does not shine at night or on cloudy days, and wind is erratic. Wind, sun and geothermal fire, within a few kilometres of the earth's surface, are rarely conveniently located and they are hampered by their scattered location, far from power stations and population centres. Until now offshore wind farms have only been established in shallow water near the shore, on massive piles drilled deep into the sea bed. Most of the so far untapped global wind resources are in deep water further out, where the winds blow stronger and with less interruption, but these have been technically unfeasible. For all these reasons it has sometimes been estimated that no more than 20 per cent of energy can be supplied from sources that need base-load backup.

There are other problems. Even offshore wind farms have been stalled by local opposition, if visible from the land, as eyesores and a danger to birds, and this has stalled their development in the US. Solar thermal power stations in desert areas require scarce water for cooling. The technology for economical deep drilling, for the hottest and most widespread sources of geothermal energy, is still being developed, as is the expertise to ensure there are no seismic disturbances.

Technological advances

The 20 per cent limit is being called into question, as progress is made in cutting costs and overcoming the obstacles. New exfoliation techniques, for example, have cut the cost of solar cells by half. Also being developed are new, more efficient shapes for wind turbine blades, some copied from the bumps on the fins of humpback whales.

The prospects, for cheaper and longer lived batteries for storage of intermittent power, has been furthered by discovery of large new sources of lithium in Bolivia. Developments in vanadium-based flow batteries (developed in Australia and providing 40 per cent of the electricity supply on King Island) allow storage of large amounts of wind power. As electric cars come on line, they will effectively provide millions of batteries that will store wind power generated at night, when other demand is low.

DeepCwind, a consortium of universities and private companies, is testing a number of designs for mooring wind turbines to platforms in deep water, off coastal cities. They plan to test a prototype in the spring of 2012, to see how far it can resist severe storms. If successful this could open up almost unlimited potential for the expansion of wind power.

New infrastructure

The most important solution to the problems of scale, location and intermittency undoubtedly lies in the construction of major new infrastructure – transnational, even transcontinental direct current (DC) power grids to collect, combine and distribute the new sources of energy. These will involve major costs and may raise issues of conflicting national interests and of security from terrorist attacks that will need to be resolved. The US doubled its wind power in 2008, overtaking Germany, but had 300,000MW of proposed wind projects waiting to connect to the grid. By mid-2009 plans were in progress for new transmission lines across the Midwest plains, and Obama's stimulus package had allotted US $11 billion to modernise the grid.

Plans were formally drawn up in January 2010, at a meeting of nine European countries, to start building an undersea electricity grid, within the next decade, to link Europe's clean energy projects around the North Sea. It would combine electricity from Britain's offshore wind projects and Germany's vast array of solar panels, and connect them also to Norway's hydro-electric power stations, which could act as a giant battery for Europe's clean energy. It could potentially also draw power from any future geothermal projects, and later could also be connected to the Euro-Mediterranean Partnership Desertec project.

Fred Pearce describes how this Desertec project, to link Europe to Saharan solar generation by undersea cables, is now making progress. A consortium of twenty major German corporations, including energy utilities E.On and RWE, engineering firm Siemens, Deutsche Bank and insurance company Munich Re, agreed in 2009 to provide €400 billion to build a raft of solar thermal power plants in North Africa. The electricity would be transmitted via twenty direct current power lines from Morocco to Spain, Algeria to France, Tunisia to Italy, Libya to Greece and Egypt to Turkey. It is envisaged that the project will meet 15 per cent of all Europe's energy needs by 2050, with a peak output of 100GW.

A grid would also help to overcome some of the problems of intermittency, by evening out differences and fluctuations over long distances and between different time zones and power sources – more solar in the day, stronger winds at night. Both geothermal and hydro-electric generation provide base-load power.

During the global financial crisis in 2008 and 2009, investment in alternative energy often stalled, and hopes that stimulus packages would be focussed here were often disappointed. However, despite many continuing problems, long-term prospects in the second decade of this century may be now looking brighter, with new technological, economic and political openings promising to converge, offering prospects of longer-term reduction or solution for many of the problems.

10

ADAPTING TO A CHANGING CLIMATE

This chapter will first discuss warning systems, defences and mutual aid arrangements, in preparation for likely climate change already in the pipeline or likely as a result of delays in mitigation. It will then look at how we might respond to the impact of storms, droughts and rising seas and the consequential shortages. Finally it will discuss the possibility of reversing the damage, by protecting and replanting trees and by seemingly more far-fetched ideas for geo-engineering that might shield the earth from heat or extract carbon dioxide from the atmosphere.

Preparing for disasters

Resilience is a word used more and more across societies worldwide as decision makers realise that predicting and controlling the future does not work and that preparing for uncertainty and surprise is vital. Some natural disasters in the last decade, such as the Indian Ocean tsunami of 2004 or the recent earthquakes in Haiti, Chile and Tibetan regions in China, have been unrelated to global warming. Others, such as Hurricane Katrina in New Orleans in 2004, or the European heatwaves and the Greek and Victorian fires of 2009, or the floods in Madeira in 2010, that poured out of the too-shallow flood channels, were clearly connected with climate change. The 2010 volcanic eruption in Iceland was not itself a climate event. However, the cloud of ash that covered Europe's air space, stranded thousands of passengers and cost the airlines billions, was due to changing wind patterns in the Arctic that prevented it dispersing, and that might affect other eruptions in a similar way in future. An invention that could have protected Europe's aircraft from the volcanic ash had been mothballed by the Australian government's CSIRO twenty years before, because short-sighted airlines had been unwilling to invest in developing it. British Petroleum's catastrophic oil eruption in the Gulf of Mexico was a side effect of the desperate search for yet more fossil fuels to keep 'business as usual' going.

There are some common lessons to be learned from such catastrophes, which are likely to be increasingly relevant as we enter an era of accelerating climate change. They have all demonstrated the extent to which we are still unprepared for the unexpected, unable to cope when it strikes, and have far too few resources available for recovery. The neglected levees of New Orleans were unable to withstand Katrina, and six years on many parts of the city are still in ruins with their inhabitants gone, and medical services remain in tatters.

We must revise our conceptions of 'reasonable' precautionary measures, and prepare in advance for as yet unprecedented events. In South Asia substantial investments in flood infrastructure were still found to be too low for the exceptionally severe monsoon flooding of 2007, which caused serious damage. It seems incredible that BP had no plans for dealing with such an undersea disaster. The Australian Royal Commission into the Victorian fires of 2009, that killed 173 people, reported a year later that the emergency system had 'fallen apart under pressure', and made wide-ranging recommendations, including the burying of power cables, more back burning, preparation of evacuation plans and fire proof shelters, and a policy of 'retreat and resettlement' of people in the worst fire prone areas.

As unfamiliar crises strike, we must learn from the experience, expect repetition or worse, and share the lessons.

There is already some progress in this direction. The death rate in Haiti was observed to be much higher than in Chile, although the intensity of the quake was similar, because Chile's buildings had been planned to resist shocks. Although, six months on, most of the population of Port au

Prince was still living in tents, and cholera became epidemic. The eventual rebuilding of Haiti is intended to emulate the example of Chile, with the help from engineers at Miyamoto International and from the international NGO Architects Without Borders, working together with local professionals.

Bangladesh was better prepared for its latest category four cyclone, in 1999, with perhaps 10,000 dead compared with 140,000 in 1991 and 500,000 in 1970. The Maldives, in the aftermath of the 2004 tsunami that made twenty of their islands uninhabitable, has relocated inhabitants to new villages in a higher position, with community halls on stilts which would provide refuge against another surge.

The Red Cross *World Disasters Report* for 2009 called for early warning systems, flood walls, flexible and drought-resistant agriculture, and provision for refugees. It concluded with a call to 'build climate risk management into all our decision-making – in our agriculture, in our water management, in our urban planning', not just days but months and years ahead of possible catastrophes. For imminent disasters, there should be early warnings and evacuation plans. For medium-term preparation, escape routes should be ready. Levees and sea walls should be sufficiently high and be well maintained, ditches and waterways to channel flood waters should be kept clear, fire breaks and the reduction of fuel loads should prevent the easy spread of wild fires. In the longer term, building and planning codes should prevent construction in vulnerable materials or places.

Social preparedness is also important. If private insurance funds are unable or unwilling to take on the new risks, they must be required to do so or governments must provide the safety net. The lack of evacuation plans for New Orleans, after Hurricane Katrina, remains a paradigm of unreadiness. During the Mexican floods in 2007, thousands refused to evacuate their unprotected homes, because of fear of looters. A study by Monalisa Chatterjee of the adaptations of slum dwellers in the aftermath of exceptional floods in Mumbai, found the key to recovery was the extent of their social networks.

Holland is probably the country with the most successful experience in wresting land from the sea, and defending its gains. Dutch law mandates that its most densely populated regions be protected from a one-in-10,000 year storm and its elected water boards, dating back to the Middle

Ages, can levy taxes to ensure the old and new dams, dykes, locks and gates are well maintained. California and Louisiana have enlisted Dutch experts to help plan for a sea level rise in San Francisco Bay, and to construct storm-surge gates in New Orleans.

In the face of predictions of even greater future sea level rises, the Dutch too are updating their strategies with a philosophy of 'controlled flooding', which might also be a model for others. Instead of ever higher barriers, they are beginning to create nature reserves, parks and playgrounds that can absorb or capture flood waters, and to include temporary water storage in all new construction. Some new houses on river banks are tethered to poles and designed to float safely when the waters rise. After eight serious droughts, Uganda is developing a national twenty-five-year infrastructure and technology plan, with a countrywide assessment of irrigation potential.

Can we also help prepare adaptation for the wildlife that we have placed under threat? Suzanne Goldberg writes that scientists have long believed that 20–30 per cent of all known species of land animals, birds and fish could become extinct as a result of climate change. Recent studies put the figure higher, at between 40 and 70 per cent. Species are already evolving by themselves, moving to more temperate habitats or warmer seas and mating earlier, to escape warming. Camille Parmesan, a leading conservation biologist at the University of Texas, but something of a heretic in her field, advocates 'assisted colonisation', to move trapped species most at risk, and preferably to work towards transplantation of entire communities of plants and animals. She proposes as a start to shift some high-altitude species to higher mountains within close range. Support is growing for protected wildlife corridors to facilitate movement.

Living with a changing climate

Human societies have each learned to live in given environmental conditions, shaped within a certain range of climatic variability. Even in the absence of catastrophic events, living with climate change will place enormous pressures on ecosystems and habitual ways of living and on providing even basic needs. There will be unfamiliar infestations of pests and diseases, well outside their prior range. In some places there will be new scarcities of water for drinking and irrigation, useable arable land or fish in seas and rivers. Wherever these are diminished, this can concern distant parts of a globalised world. The water required to produce imported food, for example, was 'embedded' in it in the exporting countries.

Could the global range of our experiences, and the internationally networked resources of science and technology, come together to find ways to adapt to the new reality? It will require challenges to ingrained habits and parochialism and to vested interests in the old ways of living and producing. It will need us to build on the new independence from capitalism's short-term profit motive that scientists have shown in the global warming issue. There would not be any pain-free solutions, but we could seek to survive without the worst forms of ecologically short-sighted destruction that got us to this pass.

The changed reality will be most directly manifest in higher average global temperatures, melting ice and saline inundations. Islands and coastal populations, including some major cities, will be immediately impacted. In Bangladesh half of the rice production areas are susceptible to the salination that would result from rising seas. Increased flooding is already cutting communities off and so a network of floating schools and libraries is being developed. In parts of Africa, nomadic herding may be better suited to the changed conditions than the sedentary agriculture which has often displaced it.

Most threats, however, will not come directly from shifts in the global parameters, but from the complex disruptions of previously established weather patterns and seasons, producing more extreme and differently located temperatures and precipitation. Floods will be heavier; African droughts will last longer; if the Gulf Stream slows Northern Europe may freeze, while Australian heat and wild fires will intensify.

In facing the unfamiliar we must learn from the experience of others in different times and places as well as systematically extending and applying existing regulations, technology and science. In Australia, where rain or river flows are declining, restrictions can be imposed on watering and irrigation uses, and water rights can be bought out and discontinued, as can the planting of thirsty crops such as cotton or sugar cane. Subsidies can encourage roof-draining rainwater tanks, such as are already found in nearly half of South Australian homes. Draining water can be captured.

In Hong Kong, grey water is harvested and supplied for toilet flushing. In many places desalination plants are being increasingly considered, although the danger they may pose to aquatic ecosystems is raising concern and opposition. Drought- and salt-resistant genetically modified crops are proposed or in the pipeline.

The International Food Policy Research Institute (IFPRI) report on *Agriculture and Climate Change* estimates that climate change will tend to reduce global agricultural production and increase food prices. They project an increase of between 10 and 20 per cent ('or more') in the number of people at risk of hunger as a consequence of climate change, mostly in sub-Saharan Africa and parts of South Asia and Central America. Adaptation of farming practices could halve this projected increase, and reform of institutions could reduce it further.

The IFPRI report notes that the effects of climate change will fall disproportionately on poor farmers around the world. Cost-effective ways to help them adapt are not easy and have long lead times, but should have a high priority. They argue that often the adaptation measures needed by poor farmers also contribute to mitigation of emissions. Funding mechanisms should recognise and support such synergies, which should benefit from the international funds available for mitigation projects. For example, conservation tillage increases soil water retention, in the face of drought, while also sequestering carbon below ground. Small-scale irrigation facilities not only conserve water, but also increase crop productivity and soil carbon. Properly managed rangelands can cope better with drought and also sequester significant amounts of carbon. Genetic manipulation of crops, to increase their resistance to drought and salt, is advocated but also contested.

Will all of this suffice, in the face not only of an unmitigated or insufficiently mitigated climate change but also of a rising global population with increasing consumption demands?

A worst case scenario – relocating the world's peoples

In February 2009 *New Scientist* published an article by Gaia Vince called 'Surviving in a warmer world', which argued that if temperatures rose by four degrees most of the planet would become uninhabitable. In such a case it would be necessary to dismantle national boundaries, abandon huge areas on either side of the equator and move many of the now 9 billion people to where the water is. Areas with increased precipitation would include Canada, Siberia, Scandinavia and newly ice-free parts of Greenland in the north, and Patagonia, Tasmania and the far North of Australia, New Zealand and perhaps newly ice-free parts of the western Antarctic coast in the south. They would need to be housed in compact, high rise cities with roof gardens, eating a mainly vegetarian diet, with no fish from the now-acidic seas and just a little goat meat. Heat- and drought-resistant crops such as potatoes would replace rice. Human waste would replace cattle dung as fertiliser. Even this would not be the end if unstoppable, runaway climate change had been set in motion.

Reversing the damage

If we fail to slash emissions in time, and climate change takes off, are there any measures that can reverse the damage or halt a seemingly unstoppable momentum? Once in the atmosphere, the additional greenhouse gases we have emitted would take many centuries to degrade of their own accord, yet the biosphere recirculates a sustainable amount of carbon dioxide continually through photosynthesis and the natural processes of plant growth and decay. If this fails or is insufficient, do the last-resort ideas for geo-engineering the planet have a chance?

Reforestation

There has been considerable attention to deforestation, said to contribute some 17 per cent of climate change emissions, and to agreements and offsetting schemes to preserve forests (see Chapter 7). Strictly speaking, measures to halt deforestation are not so much about reducing emissions as about retaining the earth's capacity to re-absorb them.

A more ambitious idea would seek to replant trees in order to reverse the climate change process, assuming that the knowledge and the regulation and monitoring capacity had been perfected. UK scientists, writing in *Nature* in 2007, argued that planting forests could absorb nine times more CO_2 over a thirty-year lifetime than the 'net avoided emissions' achieved by first generation biofuels, based on grains growing on the same land. How much potential would there be for replanting forests on a large scale? How far could we increase forest cover in a future world that was hungry and thirsty?

A promising approach is to establish trees that can serve multiple purposes. If they were periodically cut and replanted or pruned, providing ligno-cellulosics as feedstock for Gen-2 biofuels, a double advantage could accrue. Another solution is the extension of agro-forestry systems, in which the planting of trees and agriculture are systematically integrated. Agro-forestry is described as a dynamic, ecologically based natural resource management system that deliberately combines woody perennials (including fruit trees) with herbaceous crops and/or animals, either in some form of spatial arrangement or temporal sequence on the same land, with the aim of diversifying and sustaining production for increased social, economic and environmental benefit.

Researchers from the University of the Philippines Los Banos–Institute of Agro-forestry (UPLB-IAF) studied a number of diverse agro-forestry trial farms, and found that the biological interactions provided a wide range of benefits including diversified income sources, increased biological production, better water quality, optimization of capture and use of scarce rainwater, and improved habitat for both humans and wildlife. The trees could serve as live fences and woodlots and as sources of fruit, firewood, medicine, animal forage and resins, and they improved soil biota and fertility and helped provide regular employment, at the same time as absorbing and holding carbon from the atmosphere. It is thus not impossible that, with proper encouragement and support, agroforestry could be developed and applied, on a large scale, to significant areas of farmland around the world, without detracting from agricultural production.

Biochar

One proposal seeks to learn from what is being discovered about the making and use of biochar, or *tera preta*, over wide areas of the Amazon basin, for thousands of years, by pre-Columbian farmers. The carbon extracted from the atmosphere by plants normally returns there again. However, with biochar, a sealed non-polluting pyrolysis converts plant and other organic waste, including straw, into a non-biodegradable charcoal that can remain permanently in the soil, adding significantly to its fertility. The Australian CSIRO thinks the idea is promising but in need of more research on its suitability for different kinds of soil, on how much carbon might be removed and on how much waste it could reprocess. A number of pilot projects are in the pipeline.

Geo-engineering

In the frustration after the logjam in Copenhagen, more scientists are contemplating the eventual possibility of geo-engineering, to pull carbon out of the atmosphere or block the sun's heat from entering, by deliberate manipulation of the planet's environment on a massive scale. This is seen as contingency planning, a necessary evil – a bad idea whose time may yet come. In March 2010, 175 scientists, from fifteen countries met in California for a five-day conference on the issue.

Fertilisation of the oceans with iron filings could promote the growth of phytoplankton that, it is hoped, would sink permanently to the bottom, carrying carbon with them. Other methods would seek to block the sun's heat, by shooting sulphur particles into the upper atmosphere, on the model of those emitted by a major volcanic eruption like that of Mount Pinatubo.

Another proposal is for wind-driven churners to mix and cool the waters in such places as the Bay of Mexico, and thus pre-empt hurricanes.

Critics fear that such measures could interfere seriously with weather systems or an ocean ecology that we barely understand, or cause holes in the ozone layer. It is also not apparent that the effects could be controlled or directed. If used to neutralise continued rising greenhouse emissions they would be 'addictive', and could never be halted without risking immediate catastrophe. So far there has been talk but little research funding, and even advocates see it as a counsel of despair, on the agenda far down the track.

Conclusion

In the previous chapters we discussed various approaches to mitigating climate change – slowing down and then halting the process. In this chapter we have discussed what to do when we fail and if we continue to fail. In the final part of the book we will look at the prospects for success and failure. At the time of writing in August 2010, the most likely outcome seems to be a delayed and inadequate response that might, if we are lucky, be partially effective at staving off the worst. At best it may be too late for our children and grandchildren to survive unscathed, given the already accumulated and long-lived damage to the atmosphere, to the warming and acidifying deep oceans, to the lost albedo of the melted and shattered ice and to many ecosystems. However, if we start preparing our mitigation and adaptations now, we might be able to reduce the extent of harm and perhaps ensure that our great-grandchildren can emerge and reconstruct, and avoid repeating the mistakes of yesterday and today.

PART III
Who can do it?

11

FROM THE BOTTOM UP
OR THE TOP DOWN?

There is not any simple or unitary strategy for successfully promoting effective mitigation of climate change, nor for preparation and adaptation to its impacts. The issues raised are often polarising but the cleavages are not consistent, nor do they follow historically established divisions of interest or values. Initiatives at local, provincial or state, national and international levels have sometimes been more, sometimes less successful than each other, although on balance the evidence endorses the need for co-operation and mutual accommodation between levels. Every level, from the local up, can form a major delaying obstacle to new initiatives or to remedial action, if not respected and involved.

Local opposition to wind farms delayed their establishment in Britain for many years, despite government support. Conversely, US federal government inaction in the face of Hurricane Katrina inhibited effective evacuation measures at the local level. The log-jam in global negotiations, in Copenhagen in December 2009, had ripple effects all the way down. It allowed opposition to climate action to regain momentum around the world in politics and business, and for doubts about the credibility of the threat to spread among the public. It led, for example, to Australia's shelving of its proposed national emissions trading scheme, which resulted in a dramatic drop in investment in energy alternatives, as the prospects of a price on carbon receded.

Some sectors of business, such as coal and oil, have been adamant opponents of effective action. Others, including the insurance industry and alternative energy innovators, have pushed for it. Even within particular governments, parties and corporations there have been sharp divisions and oscillating policies. British Petroleum, for example, had been seeking to give real muscle to its new green logo and slogan, 'beyond petroleum', but a new chief executive officer changed the company's direction. He cut their investment in alternative energy and put new emphasis on offshore drilling – a change they must surely now be regretting!

Historical experience

Humans have long ago spread over the whole earth, and learned over centuries and generations to acclimatise themselves variously to the full range of existing conditions. Prior local experience may become less relevant in a changing climate, while useful examples may come from other places and times. Bangladesh is seeking a \$5 billion grant from the international community to improve its embankments on the model the Dutch perfected over centuries.

Excavation of ancient sites in the Caribbean, from a time when the subsidence of a tectonic plate raised sea levels locally by 5m, have shown how the people learned to stay put and adapt successfully. They constructed flimsy houses, over lagoons, on very sturdy stilts that still survive. Flooding water could pass underneath, while the people had temporary refuges prepared in caves, and could return to repair or rebuild quickly. In Belize, 2,000 years ago, the Mayans remained in areas that had became permanently flooded wetlands. They dug huge networks of drainage channels and raised their fields so that the roots sat above intruding sea water.

An environmental economist from the Australian CSIRO has described the (now abandoned) traditional Aboriginal early season patchwork-burning practices in the grasslands of northern Australia, which significantly reduced the frequency and intensity of fires later in the summer. Earlier chapters have described the old biochar practices of the pre-Columbian Amazon dwellers, and the newly developing agro-forestry techniques in the Philippines.

Different levels

Local and sub-national initiatives

There are many examples that show how local knowledge and involvement can be crucial for success. In a recent case, the small South Australian town of Whyalla announced in May 2010 that it had attracted a private consortium, together with some federal government funding, to build the largest solar installation in the country, which would provide energy to almost all its 10,000 homes within three years.

The South Australian premier claims his state

> is already a national and international leader in attracting investment in renewable energy. We have almost half of Australia's wind power, 90 per cent of its geothermal investment and more solar rooftop installations on a per household basis than any other state. We are on track to achieve our targets of having 20 per cent of electricity generation coming from renewables by 2014 and 33 per cent by 2020.

S. F. Wong describes how, in the UK in the 1990s, legal proceedings, filed by local communities opposed to wind farm development, seriously delayed many projects. In Germany, on the other hand, government regulations fostered consultation and co-ownership with local communities, which led to the emergence of many small, co-operative wind farm developers. By 2007 German wind farms had a capacity of 22,200MW, produced by 19,500 small turbines, in contrast with the UK's 2,400MW by 380 large turbines. On-shore wind farms produced over 6 per cent of Germany's total electricity.

In preparing for, and coping with disasters there is much to be gained from establishing in advance, mechanisms for co-operation. In South Australia an unprecedented heatwave, in January and February 2009, met a confused response and led to a wave of deaths among the vulnerable that 'acted as a wakeup call'. After that the state government set up an Emergency Management Council, headed by the chief of the State Emergency Service acting as a 'heat tsar'. Its task was to co-ordinate a 'whole of government' response to future episodes of extreme heat, including fire, ambulance, health, transport, water and electricity services and also on-the-ground groups like home nursing and Meals on Wheels. When the next, unseasonal, heatwave struck in November, they were ready to move together. Volunteers telephoned 17,000 pre-registered old and sick people three times a day. Young people were encouraged to door-knock elderly neighbours to check up and help. The local paper organised inspections of local primary schools, sounding the alarm about those where uncleared gutters, inflammable debris, overhanging vegetation and long grass posed a fire hazard. This time no deaths due to heat were recorded.

In 2005 the mayor of London set up an agency in partnership with the private sector, to deliver on emission reduction and low carbon targets for the city, which sees itself as very vulnerable to climate change. Despite an electoral overturn, the new mayor claims to be still focussing on investing in a decentralised infrastructure, acting as a catalyst to improve energy efficiency and opening opportunities for a low carbon economy.

The limited resources and knowledge of local initiatives may, however, present obstacles to expansion, and they can face electoral challenges from populist opponents, and the withdrawal of business funding for their campaigns. The mayors of both Montreal and Ottawa, however, were defeated by opponents promising tax cuts, after they had introduced bicycle ways and new public transport. The German wind farm expansion slowed with saturation of available on-shore sites, and more long-term prospects favoured the British. As technological development came to suit larger, offshore wind farms, the small size of the German operators prevented them taking advantage of these capital intensive but profitable opportunities. The larger British power producers, already experienced in the field, were less restricted by regulations and happy to bypass local obstacles by moving offshore. At the end of the decade the UK had plans to obtain more than a quarter of its electricity from offshore wind, wave and tidal power generators by 2020. In the United States, on the other hand, only the first offshore wind farm, off Cape

Cod, was approved at the end of the decade, after nine years of local opposition.

As unpredictable impacts of climate change, storms, droughts and floods become more frequent, it is clear that the preparations and responses of individuals and local communities cannot suffice by themselves.

Trans-local networks

While local and sub-national communities and authorities may have limited resources and knowledge, they may be able to magnify these by national and transnational networking. California, determined to take action despite opposition from the Bush federal government, initiated an alliance of 110 US and Canadian city mayors, to set up their own emissions trading scheme.

In the face of disaster, local resources are likely to be limited. More widely networked individuals and groups may offer the best prospects for rapid assistance. Local knowledge may be indispensable but it cannot foresee the changing reality of the global climate, for which the pooling of human experience and collaborative scientific study on a national or global scale are required.

Australian scientists have been mapping the coastline of the entire continent to identify locations vulnerable to rising seas, providing local governments with the confidence and knowledge to impose building restrictions on protesting owners of previously desirable sea front land. Scientists are using thermal satellite imaging of the whole world, to map wild fires and ascertain which areas will be most vulnerable. In the aftermath of the Indian Ocean tsunami of 2004 has come the establishment of an early warning system around the whole region, so that when a tidal wave first appears, those in more distant places can be warned rapidly to seek higher ground.

After half a century of growing global migration and communications, the earthquake in Haiti saw the integration of the local with the global undermined by failure at national level. Immigrant Haitian communities in North America, networking with relatives and friends around the world and back home, were able to organise substantial and well targeted relief, and to publicise the needs of the victims.

National governments

In a world of often isolated individuals and of weak local communities, effective action often requires significant support from national governments. Six months after the Haitian earthquake, reconstruction was still being blocked, apparently by a vacuum at the level of the national government, which was unable to channel the aid sent or promised by foreign governments.

There are success stories where government has been able to motivate local and private sector actions, when circumstances and public opinion were favourable. Some of the most effective outcomes have resulted from co-operation between different levels of government, together with local communities and innovating entrepreneurs. Through selective subsidies and taxes, the direct undertaking of new infrastructure projects and the imposition and enforcement of laws and regulations, national governments can seek to bypass local vested interests and the hidebound, short-term horizons of established energy companies.

In Denmark, where powerful winds are available, large businesses, willing to seek profit through long-term strategic planning, have come together with strong government regulations, subsidies and funded research that select sites, puts them out to tender and mandates local community consultation and shareholding. Responding to the oil crisis of 1973, Denmark became a pioneer in alternative energy, and is now a leader in offshore wind technology that supplies 20 per cent of the country's electricity and 11 per cent of its exports. Its large utilities have plans to increase these figures substantially and to provide fuel for the batteries of a new industry of mass produced electric cars. These projects have certainly been facilitated by the absence of big indigenous oil and auto companies, motivated to erect obstacles to change.

For many years the different Australian states have been in conflict over the distribution of the water flowing down the river Murray, with major upstream irrigators (powerful within their own states) siphoning off much of it. As the prolonged drought threatens the survival of the ecosystems of the lower reaches and lakes, moves are at last being taken, with much continuing dispute, to give the overall authority to the federal government.

Relying only on local action, where knowledge and resources may be limited, or only on distant national or global institutions, where motivation and specific understanding may be absent, can both be ineffective. We can progressively and painfully learn from immediate local experience, but we will need also to bring to bear the lessons and resources from the national level, from many parts of the world and from human history.

Conclusion

Wherever a major disaster threatens or strikes, the combined help and expertise of people from around the world is needed. In an era of rapid climate change, disasters and destructive impacts arrive in unfamiliar ways and places, inevitably finding many unready. Preparing and adapting requires co-ordinated efforts at every level, from the individual to the global. The scale of the transformation needed to halt the damage caused by emissions, and to start turning the situation round, cannot be effectively confronted without globally co-ordinated action as well. To this the next chapter will be addressed.

Fritze and Wiseman (2009) suggest that:

> The interaction of climate change policy across scales of governane can be envisaged as a series of nested 'Russian dolls' with each determining the context in which the next level operates, from the global to the local.

12

GLOBAL CONFLICT OR CO-OPERATION?

Without an enforceable global agreement on emissions reductions, the fear of free riders presents huge obstacles to effective mitigation action. Not only is it claimed the cost of such measures would give competitive advantage to rival companies and nations, but that it would also thereby undermine the environmental effectiveness of the measures, by simply shifting emissions offshore. Without the assurance of clear targets and long-term plans for a carbon price to level the playing field, investment in alternatives cannot get off the ground.

Global co-operation or conflict – a balance sheet

On 17 July 2010 the UN's outgoing chief negotiator, Yvo de Boer, was interviewed on the Australian Broadcasting Corporation programme *Lateline* by Tony Jones. Jones noted that many saw the Copenhagen conference of December 2009 as a complete disaster and he asked how profoundly the failure had set back the cause of climate change action.

De Boer replied that he thought, for those expecting a legally binding treaty, this was certainly a disappointment, but that for him it represented a significant advance. One hundred and twenty heads of state and government had come, and 127 countries had signed up to the Copenhagen

Accord. This had formulated a long-term goal and had mobilised $100 billion in finance. Since then all industrialised countries had submitted 2020 emission reduction targets, and thirty-five developing countries had proposed national action plans. These countries accounted together for 80 per cent of global greenhouse gas emissions.

Copenhagen, said de Boer, needed to deliver a broader outcome in terms of the engagement of developing countries and at the end of the day, after a very difficult negotiation, those commitments were given by China, India, Brazil and a number of others. The political commitment was very significant. Copenhagen marked a turning point in the sense that in the past, in the context of the Kyoto protocol, it had all been about negotiating targets at the international level, and Copenhagen marked the moment when the world really began to focus much more on operational frameworks within which countries make commitments.

Now you can definitely say, he continued, that those commitments are not ambitious enough. But certainly we have managed to widen the net and get a broader engagement from the international community.

President Obama, said de Boer, had invested personally in seeking out the Chinese, the Indians and the Brazilians to broker the final deal. What of course was critical for him was to get an outcome which would not make it more difficult to get legislation adopted by the US Senate. This meant getting a commitment from China that could be held up to the American people as a significant advance on the part of a developing country. De Boer thought that Obama had made progress on that front.

What de Boer saw around the world was emissions trading being recognised as an effective way of moving forward, but a good dialogue with industry would be needed in order to achieve an environmental goal without economic damage. The lack of a global agreement on mandated emissions was due to the continuing doubts of many about the potential for 'green economic growth' and about the compatibility of economic and environmental goals. This was particularly significant in developing countries where the overriding concern was economic growth and poverty eradication; two goals they were not willing to jeopardise in the context of an issue which they felt had been caused by the richer countries.

In addition, the 'Climategate' email scandal had done a lot of damage to public confidence that this was a real issue, reducing the public backing any politician needed in order to act boldly. 'We need to reinvest in building that public confidence and understanding.'

De Boer doubted if adequate emissions targets were likely in the next decade. The international scientific community had said that we needed to see a peaking of global emissions in about the next ten years, and if industrialised countries were to take their fair share of the targets to get us there they needed to reduce their emissions by 25–40 per cent. The offers on the table were nothing near to that range. The only way to get us out of that dilemma was to have a stronger economic debate, including the business community, on how to proceed in a way that was meaningful for the environment and workable for business.

Asked about the prospects for the resumed negotiations in Mexico in December, de Boer said that Copenhagen had showed the impossibility at the moment of getting major developing countries like China and India to commit to legally binding targets at an international level. What had to be done at Cancun in Mexico, at the end of this year, was focus much more on the operational frameworks.

> I believe there are real cost saving opportunities to reduce greenhouse gas emissions. I really do believe that there is a way forward to green the economy, moving into the future. We basically have to, in a dialogue between the public and the private sectors, find that way forward together. And I would rather see a broad approach at the international level than an endless debate on the level of ambition which ultimately is not agreed.
>
> *(BBC)*

What is needed

John Holdren, director of the US Office of Science and Technology Policy, has stated that: 'In my view we are already experiencing dangerous [climate change]' but that a rise in global average surface temperature of more than 2°C would be 'catastrophic'. Rajendra Chaudhaury, chair of the UN's Intergovernmental Panel on Climate Change, told a meeting of the G8 in July 2009 that, in order to limit temperature increases to no more than two degrees to 2.4°C they would have to peak global emissions no later than 2015. James Hansen of NASA, and the German government's advisory Council on Global Change, calculate that the world has a 'budget' of 750 gigatonnes of CO_2 over the next forty years to have a two out of three chance of holding temperature rises under 2°, or of just 420 gigatonnes to stay below 1.5°. At the current rate of pledges on the table, the whole of this forty-year carbon budget will be spent in the next decade. We now have five years to bring about a major change in the way we have been doing business all over the world.

Meeting this deadline now seems increasingly unlikely. In the US, carbon reduction has been blocked in the Senate, in the name of the competitive advantage it would give to China. Australia's major contribution to warming is not in the consumption levels of its small population, but in its role as the largest exporter of coal, with a quarter of the world's total. Its now abandoned emissions trading scheme included offers of 'compensation' for exporters, that would have largely negated its impact on the world price and consumption of coal. Only a global tax or impost would undermine such interests and arguments.

Before individuals, action groups, large and small businesses, markets or governments can really take cumulative and effective action towards a decarbonised future, a binding and enforceable global framework is necessary. This should include:

- Agreed short-, medium- and long-term emissions reduction targets for all. This will vary between developed and developing countries, according to different needs and resources. Funding from wealthier countries for adaptation and for leapfrog technology, would be part of the bargaining.
- A meaningful and predictably rising price on carbon (either as a simple tax or as some kind of trading scheme) to compensate for the externalities and to motivate a shift to alternatives.
- Funding for substantial international research into all the options, and a shared ownership of the outcomes.
- The co-operative initiation of major cross-border infrastructure projects, in particular direct current (DC) electricity grids and fast train lines, with access for all diverse small and large producers.

It will need to incorporate a range of trade-offs and compromises to bring everyone on board, in particular significant assistance with costs of both adaptation and mitigation in poorer countries. The conference in Copenhagen at the end of 2009 was unable to overcome conflict and mistrust. There is, however, no alternative to continued efforts towards a co-operative and fair outcome.

13
CONCLUSION

Sitting at my computer in August 2010, I do not have the temerity to write a conclusion to this book. The fearsome complexity of the inter-acting natural and social systems involved has defeated the predictive powers of many far more expert than myself. Natural scientists have found the accelerating and cascading changes unfolding beyond their expectations. Social scientists have struggled to know where our responses are going. The meetings in Kyoto in 1995 and in Bali in 2005 were salvaged at the last minute, to the surprise of many. Much was expected from the conference in Copenhagen at the end of the decade, yet it ran into the sand, and so far this has taken the wind out of the sails of action at national and regional levels as well.

The news has just come in that the Democrats have abandoned what now seemed a doomed attempt to introduce a carbon tax to the US Senate. With mid-term elections likely to increase the number of Republicans, and some Democratic senators worried about jobs in coal dependent electorates, there seems to be no chance of comprehensive climate-change legislation for years. It is unlikely that America can meet its promise at Copenhagen to cut its emissions by 17 per cent (below 2005 levels) by 2020, and other countries' promises are likely to follow suit. The Australian Labor party now has a derisory target of 5 per cent reduction by 2020, and has postponed significant legislation until after the 2013 election at the earliest. World wide, the media and politicians are paying less attention to a deadlocked issue and there is evidence that public confusion and uncertainty is again increasing.

On the other hand, while the brief window of opportunity to act in time is closing fast, the problem will not go away. The overall upward trend of global warming continues, with a changing climate increasingly expressed in the greater probability of unprecedented weather events in many parts of the world. NASA's Goddard Institute has now collated air, land and sea measurements for the first half of 2010, and found the average global temperature to be the highest ever recorded. The year is well on the way to experiencing the hottest global average ever, exceeding the previous record temperature of 1998.

Moscow has just measured 39°, its hottest day since records began 130 years ago, and is battling unfamiliar fires and drought. With its wheat crop affected, it is halting exports of grain. The polar ice continues to fracture and melt. In the largest such event in fifty years, a 260km² ice island has just broken away from Greenland. As atmospheric humidity rises and weather systems are increasingly disrupted, the rivers of Central Europe have risen to unprecedented heights; Niger and Kenya are suffering from exceptionally heavy rains after periods of severe drought; China is experiencing record-breaking mud slides.

In what is being claimed as perhaps the worst natural disaster since records began, the monsoonal flooding that is devastating Pakistan has killed thousands and affected at least 20 million. In the case of the Indian Ocean tsunami of 2004, and of the earthquake in Haiti earlier in 2010, global empathy and solidarity mobilised significant assistance. So far, the suspicions fostered by terrorism and by the war on terror have minimised any such global response to this catastrophe. John Urry's conflict-ridden and fragmenting world, as inadequate pre-emptive action fails to curb the destruction of human livelihoods and the breakdown of society, does not seem so improbably alarmist.

SOURCES AND FURTHER READING

Books and articles

Archer, D. (2009) *The Long Thaw: How Humans are Changing the Next 100,000 Years of Earth's Climate*, Princeton NJ: Princeton University Press.

Berners-Lee, M. (2010) *How Bad Are Bananas? The Carbon Footprint of Everything*, London: Green Profile.

Dawson, B. and Spannagle, M. (2009) *The Complete Guide to Climate Change*, London: Routledge.

Diamond, J. (2005) *Collapse: How Societies Choose to Fail or Survive*, London: Allen Lane.

Eggers, D. (2009) *Zeitun*, London: Hamish Hamilton.

Fritze, Jess and Wiseman, John (2009) 'Climate justice: key debates, goals and strategies' in J. Moss (ed.) *Climate Change and Social Justice*, MUP, Melbourne.

Giddens, A. (2009) *The Politics of Climate Change*, Cambridge: Polity Press.

Gore, A. (2009) *Our Choice: A Plan to Solve the Climate Crisis*, London: Bloomsbury.

Hansen, J. (2009) *Storms for my Grandchildren*, London: Bloomsbury.

Hooper, Meredith (2007) *The Ferocious Summer: Palmer's Penguins and the Warming of Antarctica*, Profile Books, London.

Lever-Tracy, C. (ed.) (2010) *Routledge Handbook of Climate Change and Society*, London: Routledge.

Morgan, G. and McCrystal, J. (2009) *Poles Apart: Beyond the Shouting, Who's Right about Climate Change?* Carlton North, Victoria: Scribe publications.

Moss, J. (ed.) (2009) *Climate Change and Social Justice*, Melbourne: Melbourne University Press.

Pearce, F. (2010) 'Earth's nine lives', *New Scientist*, 27 February: 31–35. Available online at www.newscientist.com/topic/climate-change.

Pittock, A. B. (2009) *Climate Change: The Science, Impacts and Solutions*, Collingwood, Victoria: CSIRO Publishing.

Rifkin, J. (2009) *The Empathic Civilization: The Race to Global Consciousness in a World in Crisis*, Harmondsworth: Penguin, chs 12, 13.

Ringler, C., Biswas, A. and Cline, S. (eds) (2010) *Global Change: Impacts on Water and Food Security*, New York: Springer.

Schneider, S. H. (2009) *Science as a Contact Sport: Inside the Battle to Save Earth's Climate*, Washington DC: National Geographic Society, chs 5–9.

Urry, J. (2008) 'Climate change, travel and complex futures', *British Journal of Sociology*, 59: 261–79.

Walker, G. and King, D. (2008) *The Hot Topic: How to Tackle Global Warming and Still Keep the Lights on*, London: Bloomsbury.

Websites and periodicals

Australian Broadcasting Corporation website: www.abc.net.au/news/events/climate-change/. www.abc.net.au/environment/

Blog: http://blogs.chron.com/sciguy/archives/climate_change

British Broadcasting Corporation website: www.bbc.co.uk/search/climate_ change

Campaign for emissions reductions: www.350.org

Climatic Change Journal: http://springerlink.com/content/100247/

Climate Policy Journal: www.climate-policy.com/

Environment Magazine: www.environmentmagazine.org/Archives/topic-lp/Climate-Change.html

Guardian Weekly, newspaper and website: www.guardian.co.uk/environment

James Hanson website: Continually updating climate data. www.columbia.edu/~mhs119/

International Journal of Climate Change: Impacts and Responses: http://on-climate.com/journal/

Nature: www.nature.com/nclimate

New Scientist: www.newscientist.com/topic/climate-change

INDEX